T0257112

THE PARTICIPATORY CONDITION
IN THE DIGITAL AGE

ELECTRONIC MEDIATIONS

Series Editors: N. Katherine Hayles, Peter Krapp, Rita Raley, and Samuel Weber

Founding Editor: Mark Poster

(continued on page 305)

The Participatory Condition in the Digital Age

. . . . .

Darin Barney, Gabriella Coleman,
Christine Ross, Jonathan Sterne,
and Tamar Tembeck, Editors

Electronic Mediations 51

University of Minnesota Press
Minneapolis
London

Bernard Stiegler, "The Formation of New Reason: Seven Proposals for the Renewal of Education," translated by Daniel Ross, was originally published in French as "L'École, le numérique et la société qui vient" by Bernard Stiegler, copyright Mille et une nuits, département de la Librairie Arthème Fayard, 2012.

Published by the University of Minnesota Press
111 Third Avenue South, Suite 290
Minneapolis, MN 55401-2520
http://www.upress.umn.edu

Printed in the United States of America on acid-free paper

The University of Minnesota is an equal-opportunity educator and employer.

23 22 21 20 19 18 17 16   10 9 8 7 6 5 4 3 2 1

Library of Congress Cataloging-in-Publication Data

Names: Barney, Darin, editor.
Title: The participatory condition in the digital age / Darin Barney,
    Gabriella Coleman, Christine Ross, Jonathan Sterne, and Tamar Tembeck, editors.
Description: Minneapolis : University of Minnesota Press, [2016] |
    Series: Electronic mediations ; 51 | Includes bibliographical references and index.
Identifiers: LCCN 2016003045 | ISBN 978-0-8166-9770-0 (hc) |
    ISBN 978-0-8166-9771-7 (pb)
Subjects: LCSH: Participation. | Social participation. | Political participation. |
    Internet—Social aspects.
Classification: LCC HM771 .P367 2016 | DDC 302/.14—dc23
LC record available at http://lccn.loc.gov/2016003045

Contents

III. Participation under Surveillance

IV. Participation and Aisthesis

The Participatory Condition

An Introduction

*Darin Barney, Gabriella Coleman, Christine Ross,
Jonathan Sterne, and Tamar Tembeck*

THE PARTICIPATORY CONDITION names the situation in which participation—being involved in doing something and taking part in something with others—has become both environmental (a state of affairs) and normative (a binding principle of right action).[1] Participation is the general condition in which many of us live or seek to live (though, to be sure, not all of us, and not all in the same way). It has become a contextual feature of everyday life in the liberal, capitalist, and technological societies of the contemporary West. It could be argued that there is no place or time in human history where and when people did not "participate" by living together and acting in their world. Participation is, after all, the relational principle of being together in any civilization, society, or community. However, the fact that we have always necessarily participated does not mean that we have always lived under the participatory condition. What is distinctive about the present conjuncture is the degree and extent to which the everyday social, economic, cultural, and political activities that comprise simply being in the world have been thematized and organized around the priority of participation *as such*.

The generalization of participation is concomitant with the development and popularization of so-called digital media, especially personal computers, networking technologies, the Internet, the World Wide Web, and video games. These media allow a growing number of people to access, modify, store, circulate, and share media content. The expansion of participation as a relational possibility has become manifest in the variety of fields media participation embraces, including: participatory democracy (from representative to direct democracies and on to the development of collaborative commons and the Occupy movements), citizen journalism, social media communication, the digital humanities, digital design, smart cities, gaming, and collaborative art. But what does it mean to participate? How and why is it that we believe that we now participate more? What are the main features of participation today? And why has it become so important?

Participation is not only a concept and a set of practices; fundamentally, it is the promise and expectation that one can be actively involved with others in decision-making processes that affect the evolution of social bonds, communities, systems of knowledge, and organizations, as well as politics and culture. Tied to this promise and belief, as well as to the structures of the media technologies (Internet forums, blogs, wikis, podcasts, smartphones, etc.) that appear to facilitate increased participation, are the possibilities of communication linked to social change. But while possibilities represent desire, they can also be understood as rhetoric, as a set of empty habits, or as failed opportunities. This tension—between the promises and impasses of participation, its hopes and disappointments, its illusions and recuperations—is at the forefront of recent social, cultural, and political assessments of participation in relation to new media. Attending to this tension, *The Participatory Condition* critically probes the purported participatory nature attributed to media, and unearths other forms of participation that might be obscured by excessive promises of digital utopias.

Henry Jenkins's work on "participatory culture" helps to clarify the specificities of the present conjuncture. Jenkins first coined the term in 1992 to describe the cultural production and social interactions of fan communities.[2] The term has since then evolved in his coauthored publications, namely *Convergence Culture* (2006), *Confronting the Challenges of Participatory Culture* (2009), and *Spreadable Media* (2013)—studies that account for the relations between the development of participatory culture, the evolution of new media technologies, the expansion of the various communities invested in media production and circulation, and the decentralization of decision-making processes. Key to Jenkins's understanding of participatory culture is its articulation as not only emergent but also expansive, owing largely to the "spreadability" of emerging media—a paradigm that "assumes that anything worth hearing will circulate through any and all available channels."[3] We would agree, adding that the expansive quality of participation demands a shift in terminology. The proposition of this book is that the normative and environmental qualities of participation that Jenkins and others have assigned to culture have now been generalized across multiple social domains such that it becomes possible—perhaps even necessary—to start talking and thinking about a "participatory condition" whose operations and implications exceed the boundaries of a single culture.

This volume has three main objectives. First, it collects the work of key scholars of participation and new media, across a wide range of disciplines, in order to disentangle the tensions, contradictions, and potentialities of new media participation. Second, each of its essays seeks to assess the role of new media in the development of a relational possibility—participation—whose expansion has become so large that it represents the very condition of our contemporaneity. Third, it affirms that, while in recent years the term participation has come to be associated with digital media and the social web (or "Web 2.0") in particular, the concept has a long history that predates and informs the digital age. The contemporary participatory condition relies upon a number of historical "preconditions" across the fields of politics, art, and media. This complex history includes a range of ideas, practices, and artifacts that cannot be reduced to, nor wholly accounted for by, technological changes alone. While a detailed look into the many interrelated layers of that history is beyond the scope of this introduction, our brief examination of these preconditions helps to better situate contemporary participatory practices and their evolution into what we identify as a *condition*.

Participation as Interpellation

Across a broad range of social domains, our expectations to participate are matched with expectations that we will participate. Participation has become a measure of the quality of our social situations and interactions, and has come to stand in for virtues that, under other conditions, might have names like equality, justice, fairness, community, or freedom. Participation is normal; a lack of participation seems suspicious, strange, and disappointing—an impoverishment of democratic forms of citizenship which normally involve "equality as participation."[4] Participation has become a tremendously valuable social, political, and economic resource. In this sense, the participatory condition names a particular instance of what Louis Althusser described as interpellation, the process whereby we become the subjects we are by responding to the hail of ideological formations that structure our social environments. In his 1970 essay "Ideology and Ideo-logical State Apparatuses (Notes towards an Investigation)," Althusser describes the primary scene of interpellation as the hailing and hearing of a lawful exclamation: A police officer shouts out in public, "Hey! You there," prompting an individual to turn around, whereupon, "by this mere one-hundred-and-eight-degree physical conversion, he becomes a *subject.*"[5] In the present condition, we are hailed as participants by multi-ple elements of our environment across the domains of culture, politics, and social life. Recognizing ourselves in that hail, we act accordingly: We participate.

That participation has evolved into a leading mode of subjective inter-pellation in the contemporary period is the central assumption of this book. Participation is one of the most prominent means by which indi-viduals and publics (at least in the contemporary West) become subjects and inscribe themselves in the social order. We participate in the process of becoming participatory subjects, but an element of contingency per-sists in any situation where human agency is at play. This is another sense in which participation is conditional: What it means depends upon what we become as participatory subjects, and this is not simply given in advance. The participatory condition is what we live with, with all the constraints and possibilities that living-with implies.

Participation is not a quality added to some other thing or activity, not one hailing process among others, but a condition that is constitutive of the social itself. The present volume explores the multiple, complex, and sometimes contradictory elements of the participatory condition across the domains of politics, art, and social life. Considered together, the essays in this volume specify what makes participation a condition today—what makes it such an interpellant force: its generalization (it extends in a grow-ing range of societal fields and has become an umbrella for diverse prac-tices and phenomena); its strong compatibility with neoliberalism as a political economy; and its refraction in contemporary media environ-ments. Moreover, as the collected essays suggest, it becomes increasingly clear that contemporary participation has become a *pharmakon* of sorts, to borrow one of the key concepts from Bernard Stiegler's philosophy of technology: both a poison and a remedy, a benefit and a problem, a prom-ise of emancipation as well as a form of subjection.

The participatory condition requires us to think beyond accounts that would simply equate it with the rise of digital technologies. It goes without saying that the participatory condition is intimately bound up with these technologies, and their extension into and across multiple domains of social, political, economic, and cultural practice. Scholarship that has focused on the correlation between participation and digital media—namely, Jenkins et al.'s work on "participatory culture" (1992–2013), Nico Carpentier's *Media and Participation: A Site of Ideological-Democratic Struggle* (2011), and Aaron Delwiche and Jennifer Jacobs Henderson's *The Participatory Cultures Handbook* (2013)—recognizes the central role of the digital in the expansion of participatory cultures, but also stipulates that the latter are not determined by and do not result from the development of the former. The essays in this volume confirm this sense that new media are a necessary but insufficient condition of the broader participatory condition. In *Confronting the Challenges of Participatory Culture* (2009), Jenkins et al. characterize communication technologies as one element within the complex ecology of participation: "Rather than dealing with each technology in isolation, we would do better to take an ecological approach, thinking about the interrelationship among different communication technologies, the cultural communities that grow up around them, and the activities they support. Media systems consist of communication technologies and the social, cultural, legal, political, and economic institutions, practices, and protocols that shape and surround them."[6] The interdisciplinary essays collected here attempt to apprehend the participatory condition in exactly such an ecological manner. Another characteristic of ecologies is that they exist in time—they develop from something and into something. From what, if not *only* media (but yes, *also* from media) has the participatory condition developed? Before speculating on where the participatory condition might be going, it is necessary to consider where it came from.

A Political History of Participation

The identification of politics with participation has a long history in the Western tradition, a history that has prepared us to both expect and accept participation as definitive of political experience itself. This long history has primed us to receive the hail of the participatory condition. In the classic Aristotelian definition, a citizen is one who participates, specifically, "in the administration of justice and the holding of office."[7] By contrast, modern understandings of citizenship emphasize membership or belonging, the status of being a recognized part of a political community. From late eighteenth-century modernity on, participation has referred to citizenship in both senses: To participate is to be a part and to do your part. Already this signals a structuring tension at the core of citizenship, as it is understood in the West, which can be figured as the nonidentity between participation as acting and participation as belonging. Put simply, not everyone who participates in a political community by belonging to it participates in that community by actively taking part in its political life. Many who belong would prefer not to participate, and liberalism makes this their right (even as emerging technologies hollow out this right by making participation compulsory).[8]

At the same time, the actual history of liberal democratic societies is one in which active participation in political life for some members of the community rests upon the structural exclusion of other members of the community from participation in political institutions and the public sphere. The material reality of a class of *participants-who-cannot-participate* was already present in Aristotle's classic formulation of citizenship, whereby slaves and women who "belonged" to the household—and, by extension, to the *polis*—were excluded from "the administration of justice and the holding of office" *as a condition of the possibility of participation* by Greek male citizens. Women and slaves were, to borrow the phrasing of Jacques Rancière, the "part that has no part."[9] Participation as membership does not guarantee that one qualifies as a participant in the more active sense, a reality that persists materially even as formal rights to participate are extended to more and different classes of citizens. Indeed, as Rancière observes, much of the substance and history of politics can be attributed to the never-ending struggles of participants-who-cannot-participate—that is, the part that has no part—to rearrange the distribution of parts.

This structural tension has placed participation at the center of the Western political imaginary. On the one hand, what is understood to define liberal democratic societies is their institutionalization of political participation by citizens in forms that include: constitutional protection of free speech, association, and assembly; citizen suffrage and eligibility for office; responsible government and elected legislatures; political parties; citizen initiatives, recall, and referenda; stakeholder consultation and public regulatory hearings; and mandated public access to state information. On the other hand, expectations regarding participation have also motivated critiques of these very institutions as insufficiently democratic, whether because various classes of people have been denied access to them on account of not being recognized as qualified participants, because the mechanisms of representation make for a political life that is insufficiently participatory, or because a robust democratic life demands extension of participatory principles beyond the limits of institutional politics into those social and economic domains in which power is actually organized. In each case, participation is confirmed as central to what politics actually is. Thus, participation has been a core value that traverses several diverse and sometimes contending categories of political thought and practice in the West.[10]

Participation has also figured centrally in twentieth-century philosophical accounts that locate politics not in the formal institutions of democratic government but rather in the informal settings and interactions of the democratic public sphere. In Hannah Arendt's account of the Athenian *polis,* one becomes political not by occupying office but by participating in public life, appearing before others as an equal, and committing to action in speech.[11] In Arendt's view, such participation is not validated instrumentally by the ends it achieves but is sufficient unto itself for the realization of a distinctive human excellence: Participation is its own reward. In his influential account of the modern, liberal democratic public sphere, Jürgen Habermas casts politics as participation by private persons (i.e., nonholders of public office) in the formation of public opinion through rational-critical debate in the public sphere.[12] Here, participation has the specific, democratic purpose of justifying shared norms and testing the legitimacy of state authority. Such theories of the public sphere exhibit two important tendencies that bear on the place of participation in the Western political imagination. The first is an explicit figuring of participation in terms of communication and the privileging of speech, rational argument, and dialogue as its definitive modes—against which other modes of communication and forms of participation are either reduced to analogues or condemned as irrational, deformed, or regressive. The second is a preoccupation with mass mediation as either enabling or corrupting the sort of communication required to fulfill the political promise of participation. Both tendencies figure participation as something that needs to be defended from damage or substitution by the many diminished forms that threaten its integrity, and as something whose prospects and reach might be enhanced by media technologies that improve the quality and quantity of horizontal communicative interaction between citizens in multiple and expanded public spheres. The participatory condition is definitely marked by these anxieties and these hopes, even as the activities that characterize it far exceed the somewhat limited frame of rational public debate concerning common affairs or the legitimacy of state authority.[13]

As prospects for a revolutionary transformation of capitalism began to dim in the 1960s, New Left thinkers and movements of the post-1968 generation turned to advocating more direct and participatory modes of democratic practice—extended across an expanding range of social, economic, and political domains—as a way to combat inequalities stemming from categorization by class, race, and sex.[14] Even contemporary anarchist and autonomist movements that promote exodus from participatory engagement with state forms and institutions considered illegitimate and unredeemable enact their abstention through highly participatory modes of political organization, hopeful that these might sustain an alternative, postcapitalist future.[15] Thus Jenkins et al. describe the origins of participatory culture in twentieth- and twenty-first-century struggles by grassroots communities and countercultural movements that fought and continue to fight "to gain greater control over the means of cultural production and circulation."[16] At the same time, the declining character of participation in public life—figured variously as occasioned by alienation, massification, atomization, privatization, bureaucratization, depoliticization, civic illiteracy, apathy, and a deficit of social capital—has been a dominant theme in Western social criticism and social science.[17]

Linking these opposed yet interdependent currents is a steadily evolving culture of self-realization and self-fulfillment that relates to participation somewhat ambiguously.[18] On the one hand, the purported individualism of this "realization of the self" ethos can be construed as a cause or a symptom of the alleged withdrawal from participation in collective, civic life that communitarian critics have diagnosed as a pathology of advanced market liberalism.[19] On the other hand, these same orientations can be credited with driving appetites for expanded opportunities to participate in a diverse range of social settings and interactions, as a means of expressing and affirming the selves we are increasingly encouraged to cultivate.[20] This raises the question of what counts as politically significant participation. For, despite what appears to be an unprecedented range of opportunities for individuals to participate in activities that seem to comport with longstanding ideas about what constitutes political action—gathering and publicizing information, expressing opinions, debating and deliberating with others, signaling preferences, making choices, witnessing events, and organizing collective action—it is not at all clear that the participatory condition marked by all this activity is actually one in which the quality, intensity, or efficacy of political experience is significantly greater, or more democratic (in the substantive sense of a more equal distribution of power and resources), than it was before participation became a routinized part of most every aspect of social life.[21]

An Art History of Participation

The questions pertaining to what constitutes effective participation in the field of politics extend into the areas of art and culture. Because of its consubstantiality with democracy, the aesthetic regime initiated in the late eighteenth century can be considered as the regime that has prepared and still prepares us for the participatory condition. Following Rancière's definition, the aesthetic regime is a regulated system of visibility and invisibility in art, as well as "a mode of interpretative discourse that itself belongs to the forms" of that system, whose main novelty is its principle of equality.[22] In postrevolutionary Europe, art ceased to exclusively represent the values of its rulers (the Church, the monarchy, the aristocracy, etc.) and increasingly became the manifestation of a sensibility—a combination of perceptions, sensations, and interpretations—toward events and objects of ordinary life. That new sensibility broke with the established hierarchy of genres and accelerated the constitution of an undifferentiated public. From then on, anything—the mundane, the unexceptional, the unidealized, the ugly, the mass cultural, the uncanny, the *informe,* and the excluded—could potentially be performed, depicted, and circulated in artistic practices. Even today, this "anything-ness" is a constant source of discontent—a discontent as old as aesthetics itself. How can anything be art? And how can art, as it increasingly attempts to bridge aesthetics and life, propose *new* forms of life? Yet it is precisely this anything-ness that presupposes and reinforces a democratic deployment of aesthetics, and that ultimately confirms the possibility of equal participation by all actors (artists, spectators, curators, etc.) in the aesthetic regime.

Underlying the rise of the participatory condition is the belief that "cultural participation" represents the full expression of the rights and capacities of human beings in democratic societies—a democratic underpinning promoted by the aesthetic regime from the start. It remains manifest today in Article 27 of the Universal Declaration of Human Rights, which states that: "Everyone has the right to freely participate in the cultural life of the community, to enjoy the arts and to share in scientific advancement and its benefits."[23] A democratic government is understood to be one that promotes the access and inclusion of its citizens in cultural life, supporting them in the full exercise of citizenship. Cultural participation effectively defines what "we" refer to as "our" culture, so in this regard, it is performative at once of our sense of belonging and of the state or community's fulfillment of its democratic ideals.

A number of participatory art practices are guided, in principle, by the promise of equality that the idea of participation seems to carry: from the historical avant-gardes (Dada, Constructivism, and Surrealism) to postwar happenings, installation art, relational aesthetics, community art, and collaborative art. Across disciplines, specific frameworks have been proposed to orchestrate collaborations between diverse publics in the hopes of giving flesh to "the possible" that is roused in the very idea of participation.[24] Of the better-known attempts at participation in the live arts in the twentieth century, we can count the modernist avant-garde's calls for provocative awakenings of the bourgeois audience;[25] the development of non-Aristotelian dramatic techniques, or epic theatre, by Erwin Piscator and most famously Bertolt Brecht, who sought to rouse the audience's critical faculties through processes of defamiliarization;[26] Augusto Boal's elaboration of methods to interpellate the "spect-actor" in more overt manners;[27] and subsequent forms of "postdramatic theatre" that straddle the traditions of the visual and performing arts.[28] In most of these practices, the exploration of emerging media (from novel forms of performance, to innovative investigations of sound, music, movement, voice, and light, and the aesthetic exploration of kinetic and cybernetic techniques, painting, film, photography, video, and the Internet) plays a significant role in the renewal of participatory aesthetics. Yet, just as participation alone is not a guarantor of the renewal of sensibilities, neither is an aesthetic renewal solely guaranteed by the integration of emerging media. The recent development of post-Internet art, which uses nondigital media to reflect upon the digital age, is an indication that art can sometimes benefit from media anachronisms.

While the names attributed to participatory practices have varied in the history of contemporary art, scholars have developed more or less stringent terminologies and criteria to identify the nature, scope, and operational parameters of participatory practices.[29] Among them, Nicolas Bourriaud's *Relational Aesthetics* and Grant Kester's writings on "dialogical art" are the most prominent examples of contemporary scholarship that sees the artistic processes of participation as primarily characterized by dialogue, exchange, and interpersonal connection; their positions, however, bare the unspoken assumption that all participants have the freedom to meet on equal grounds. For Anna Dezeuze, participatory aesthetics have the potential to extend into the political realm. "The do-it-yourself artwork," she writes, "can also serve as a catalyst for change, whether through self-consciousness and self-transformation, or through social interactions and exchanges. Participatory works are often premised on the belief that participation will encourage individuals and groups to take control of their own social and political existence," either "by offering alternative models for social or political participation" or "by acting as a means to empower participants."[30] Beryl Graham adds to this discourse a functional distinction between participatory and interactive practices as they are deployed specifically in the context of new media art: Whereas in "participatory art" participants are expected to *produce* content themselves, in "interactive art" content that was originally developed by an artist is only *refashioned* or reorganized by participant-users, thereby suggesting differing depths of creative engagement.[31]

But while participation is at times equated with the possibility of inner or outer change, its transformational value is not a given. Claire Bishop contends that participatory works operate as platforms through which antagonistic relations ought to unfold. Without antagonism, the participatory aesthetic loses its political potential—its capacity to generate new forms and, in so doing, question the social status quo. Antagonism is the means through which "the vicissitudes of collaborative authorship and spectatorship" and the merger of participation with cultural industries and spectacle can be defeated.[32] At times, participation simply functions as entertainment. This evolution follows the notion of cultural participation as described in the Universal Declaration of Human Rights, where participation includes "enjoying" culture, attending exhibitions, and being present in sports arenas or concert halls. Such an understanding of participation increasingly informs business models in the cultural sphere, as is the case with the new, audience-friendly museology by which museums have been securing their economic survival over the last two decades or so.[33] In some cases, cultural participation is forced; in others, it remains invisible. Like the new museology, art practices since the avant-gardes have often involved lay publics, who may or may not be informed of their planned participation, or even be willing to take part in the artistic event, in contrast to other forms of participatory art that build upon the consensual meetings of individuals or self-defined "communities."[34]

Even Kester, a strong supporter of collaborative practices in art, has cautioned that "a specter haunts th[e] utopian vision" of participation in the context of aesthetic experience.[35] He reminds us that modern and contemporary art's emancipatory potential has always been under threat— whether through the encroachment of mass culture and market forces or the cooptation of art as a tool for propaganda.[36] With the rise of art as participatory social practice at the turn of the twenty-first century, we risk finding ourselves in a similar position once again. In order to ensure its survival, art must follow the rules of the market or those of the state, whose cultural policies increasingly demand the justification of public support in the form of a foreseen *social* return on investment. In the current conjuncture, the emancipatory potential of aesthetic experience remains entangled with the political trappings of participation: With an art oriented toward social practice, there is always the risk of moving "from the *action of fiction*" in the creative process "to the *fiction of action*" in the world.[37]

A Media History of Participation

At least in the culture the authors of this introduction inhabit, participation is more prevalent than ever before, and more deeply tied to media. But as with the historical precedents of contemporary participatory politics and art, the media logics we are now living have deep historical roots. Each media age restages a century-long conversation about the possibilities, problems, and peculiarities of participation. Alongside its conceptual relatives—interaction, dialogue, and engagement—participation has been a major axis by which media have been heralded, described, evaluated, and criticized. Communication via media as the basis of participation is an ancient theme—as John Durham Peters notes, Plato's complaint about writing and painting was that they did not talk back; they were poor participants in dialogue.

Communicative participation took on a more explicitly political stake in the modern era. Alongside Habermas's eighteenth- and nineteenth-century coffee shops, the postal roads of the revolutionary era in the United

States were imagined as cultural infrastructures of participation, where mail and newspaper circulation was said to make possible a democratic citizenry, a trope later repeated regarding telegraphy and electricity.[38] If hopes for participation through media were connected to hopes for greater democratization, anxieties about too much participation were closely connected to issues of power and control. British and American moral panics over cheap newspapers and novels during the 1830s were tied to anxieties about the newfound power of a literate working class. We find similar anxieties around the circulation of *David Walker's Appeal* (1829)—perhaps the first document to address a black American public—where greater participation in a print public was seen as a direct challenge to white supremacy.[39] Women's participation in modern cultural and political public spheres follows a similar political topography, where the salon and the novel, and eventually public speech, print culture, and commerce, challenged patriarchal relations. Yet women's participation could also be tied to continued subjugation, as in the Victorian middle-class housewife's connection to the piano as a form of domestic entertainment.[40]

The era of technical media, to use Friedrich Kittler's term for the electrical and mechanical devices that emerged in the late nineteenth century, introduced a new range of sensorial and political controversies around participation. A crucial set of concerns grew around the distraction and disengagement effects of media—the ways in which they construct (or fail to construct) nonparticipatory participants. Composer John Philip Sousa's famous lament over the loss of live music in the age of sound recording—"What will become of the national throat?"—captured a more general intellectual anxiety that technical media would make their audiences passive consumers where before they had been active cultural producers. The retrospection was indeed expressed and perhaps experienced but, insofar as anxieties also circulated around *too much* participation, it was somewhat of a romantic recall.

The idea of media as generators of passivity remained a powerful thread in twentieth-century media theory. The concerns about media—that they short-circuited participation, or provided false or inauthentic modes of participation—were tied to more general anxieties about twentieth-century modernity. In his 1907 book *The Philosophy of Money,* sociologist Georg Simmel argued that the money economy required its subjects to become more calculating, to reduce everything to a standard of utility and exchange, and therefore tended to dull judgment and blunt aesthetic sensibilities, leading to a calculating, affectless subject who replaced authentic feeling with the search for temporary and replaceable sensations.[41] Simmel's theory of participation is thus double-edged: Successful participation in the money economy, which was obligatory for the modern subject, led to subjects' inability to participate authentically in their own cultural or psychological lives.[42] Of course, distraction did not always take on a negative valence. Most famously, Walter Benjamin wrote that the condition of cinema would lead the working classes to be able to apprehend themselves in their totality, echoing Georg Lukacs's theory of reification, where class consciousness—and the collective action that would spring from it—was only possible after capitalism objectified social relations, rendering them sensible. The Situationists' practices of *détournement* (the deviation of a previous media work, whose meaning is antagonistic to the original) also built on Lukacs's account of consciousness and action— explicitly in Guy Debord's *Society of the Spectacle* and implicitly in their attempts to turn urban distraction and alienation against itself.[43]

Another strand of theory argued that media intensified modes of participation. In his 1909 *Social Organization,* sociologist Charles Cooley identified four major characteristics of modern media that were so transformative as to "constitute a new epoch": expressiveness, permanence of record ("or the overcoming of time"), swiftness ("or the overcoming of space"), and diffusion ("or access to all classes of men").[44] For Cooley, these changes touched every aspect of modern life and led to monumental new possibilities for participation: "Never, certainly, were great masses of men so rapidly rising to higher levels as now," he wrote optimistically of mass education and public opinion. "The enlargement [afforded by new media such as the mass daily newspaper] affects not only thought but feeling, favoring the growth of a sense of common humanity, of moral unity, between nations, races and classes. Among members of a communicating whole feeling may not always be friendly, but it must be, in a sense, sympathetic, involving some consciousness of the other's point of view."[45] For Cooley, media participation enables new forms of collective feeling and action, and in his writings we find a waypoint, remarkable in a roughly two-hundred-year span, where a single set of claims is repeatedly attached to very different circumstances. Echoes of his claims can be found in twentieth-century assessments of telephony, broadcasting, satellite communication, computers, and the Internet. They were also alive in the cultural programs attached to early radio and television programming, which often followed explicitly nationalist agendas.[46]

Policy and engineering also played a role in twentieth-century debates around the distractive and potentially nonparticipatory impact of media, as well as their emancipative possibilities. Policy rulings throughout the 1910s and 1920s restricted access to radio, which had been a many-to-many medium. This changed by the 1930s, with radio largely becoming a one-to-many broadcast medium (with some notable exceptions). Engineers built technologies that conformed to policies once they were in place, but they also developed their own aesthetic theories and practices of participation and operationalized them in new media. In the 1930s, working on experimental versions of television, engineers assumed that the domestic audience would be watching while doing other things—to this day the conventions of the television soundtrack are designed to call the audience back to the television set, in case it is doing something else. The designs of portable, public, and low-definition audio and video media, from color television to transistor radios, Walkmen, MP3 and YouTube codecs, and Twitter apps for computer desktops, also assume an audience that may or may not be paying full attention, and allow for multiple levels of aesthetic engagement—a quality of participation that flows from rapt attention and engagement to rapid switching of attention, distance, and ambiance.

The twentieth-century participatory/nonparticipatory anxieties that sustained debates around the role of media in politics, culture, communication, cognition, sensoriality, policymaking, and engineering became more urgently ambivalent when set against the darker parts of what Alain Badiou has called the "short century," beginning with World War I and closing with the collapse of the USSR. Following World War I, Walter Lippmann argued for a class of experts to manage public opinion, to keep an unruly population on the proper course. The idea of propaganda captured the imaginations of militaries, governments, radicals, and critics for the next thirty-odd years. In a propagandistic world, involvement in media leads inexorably to mass mentalities—in fantasies and research around propaganda. Some scholars imagined media participation as leading to ideological injection, while others saw participation as a kind of entanglement where audiences came to move collectively, but dangerously. The legacies of fascism and a range of other authoritarianisms follow closely here. During the 1940s, Allied intellectuals worried about the Nazi propaganda machine. But the idea that media participation is the basis of a polity is still with us today, seen in NATO's bombing of broadcast infrastructure during its air raids on Serbia and the use of radio broadcasts as evidence of incitement to genocide in the trials that followed the Rwandan genocide.[47]

Understanding the Participatory Condition

Historical preconditions, including those outlined above in the areas of politics, art, and media, prepare us for the hail of the contemporary participatory condition. Modernity, especially the development of democracy it promotes, shapes itself around the promise of participation as well as the anxieties, power struggles, ambivalences, and failures that accompany such a promise. Today, the popularization of digital media reactualizes the participatory thrust of modernity across the realms of politics, art, and media, as well as beyond. The following question must then be raised: In what ways and to what extent have digital media in particular become a structuring feature of the contemporary participatory condition? This is the main question addressed in the contributions to this volume. The answers to that question are articulated within the confines of specific disciplines, including political science, sociology, communication studies, anthropology, law, philosophy, design, museology, and art. But, in light of the commonality of some of their deliberations and findings, these contributions also reach beyond their respective disciplines. An interdisciplinary assessment of the participatory condition is essential insofar as participatory culture has expanded across all societal fields.

The essays collected here also share a common concern: the need to be critical about the participatory condition. The common conclusion is that, although certain attributes of digital media facilitate participation, these attributes alone do not encompass the possibilities, promises, or deceptions of participatory practices. Rather, digital media offer environments that are ripe for the unfolding of the participatory condition. As anthropologist Tim Ingold has argued in his study of materiality, the properties of materials "cannot be identified as fixed, essential attributes of things, but are rather processual and relational."[48] Participants in new media environments (engineers, policymakers, investors, branders, employers, users, workers, thinkers, hackers, activists, players, dreamers, propagandists, educators, artists, and so on) shape the media as they are being shaped through them. These essays, therefore, all aim to analyze what, why, how, and when participation takes place within the new media environments of the contemporary participatory condition.

Although the participatory logic of contemporary media is indebted to the past, current conversations around participation have been articulated through new vocabularies and under distinct circumstances. A gloss of dominant terms used to convey the nature of online sharing—ad hoc, the commons, peer-to-peer, prosumer, user-driven innovation, spontaneity, creativity, empowerment, crowd-sourcing, and especially openness—bolsters a now-entrenched notion that the Internet is ideal for "organization without organizations," to cite the subtitle of Clay Shirky's influential book *Here Comes Everybody*, a de facto bible for this sort of thinking.[49] What is ostensibly distinct today is how the Internet allows humans to bypass institutions and hierarchies while encouraging direct connection and participation. As the anthropological record attests, humans—across time and in different societies—have engaged in radically distinct forms of participation; why would the manifestation of online participation express itself through just a singular format? As it turns out, it does not. Bart Cammaerts's essay in this volume clearly shows that information and communications technologies (ICTs) can facilitate mutual cooperation—the sharing of material goods, services, skills, and knowledge—between citizens, but the social ties consolidated by online cooperation is extremely varied, ranging from weak to strong, manifest to latent, and enduring to ephemeral. Moreover, the notion that online participation is incommensurable with the organizational styles of traditional institutions has proven to be spectacularly false. Many of the more lasting forms of participation—free and open software projects, Wikipedia, radical technology collectives, and crisis mapping groups such as Ushadi—were once informal or driven by charismatic authority but have, over time, routinized, to borrow from Weberian terminology.[50]

Nevertheless, the vision of digital participation as coterminous with flexible, nonhierarchal, and extrainstitutional endeavors continues to grip the public imagination. Recently, a cohort of theorists have insisted that we should bring more categories to bear in order to evaluate claims about novelty, political effects, and cultural significance: "Without a guide to identifying differences in participation," writes media studies scholar Adam Fish, "all forms look the same, and every instance confirms a theory rather than testing it. A field guide would allow one to observe, compare and contrast forms of participation; to ask when and where different forms occur; . . . to ask what forms of participation are emerging, what forms are going extinct, and with what consequences?"[51] In light of this, the essays in this volume help us to evaluate the place of new media in the participatory condition along four axes—politics, openness, surveillance, and aisthesis— helping us to better situate the common workings of the participatory condition across distinct fields.

1. *Politics.* In his recent study of media and participation, Carpentier focuses on the relationship between participation and decision making, and notes that the meaning of participation varies across the several fields to which this relationship pertains.[52] Notwithstanding this variety, Carpentier finds that what is at issue in these relationships remains relatively consistent across their diversity: a struggle over the distribution of power, and whether it tends toward or away from equality. The chapters in this section of the book, including Carpentier's own, all reflect on this core attribute of the relationship between participation and politics, even in cases that are not organized primarily around decision-making processes. The struggle for equality haunts the politics of participation.

The essays examining the political ramifications of the participatory condition in this volume reveal how participation operates as a promise of democratic emancipation, one that is only contingently—not necessarily—linked to egalitarian intentions or outcomes. For example, when participation takes the form of uncompensated labor that generates value for powerful corporations (Trebor Scholz), or mere consultation that applies a veneer of legitimacy to elite decision making (Carpentier), or accumulated data that enables the expansion of commercial and governmental surveillance (Julie Cohen, Mark Andrejevic), it is no longer clear that participation equates with democracy or equality. But, in line with early work on cyber-feminism arguing that through new media forms of participation new gender identities and relations would emerge, and in line with Henry Jenkins's work, some essays are more confident in their analyses of participation in relation to social and political change.[53] There are instances where media participation has indeed played a significant role in effecting such change.

One can find that potentiality in discussions of open source software movements (Alessandro Delfanti and Salvatore Iaconesi), Twitter's role in the Arab Spring (Jillian York), Tumblr's role in Occupy Wall Street (Cayley Sorochan), the role of participatory design in augmentative communications (Graham Pullin), and the promises of engaging university undergraduates in multimodal forms of education (Stiegler).[54] But even these more positive accounts agree on the insufficiency of participatory media alone in the aspiration for political or social transformation.

2. *Openness.* The stakes for insisting on a more diverse taxonomy to assess the participatory potential in new media environments can be clarified by probing the interrelations between expertise, openness, and institutions. If questions pertaining to expertise and institutions are rarely applied in attempts to understand participatory collectives online, openness, on the other hand, has been one of the privileged terms used by practitioners and commentators alike to describe many online participatory projects.[55] Popularized by the success of open source software development, its heritage lies in the modern cultures of institutional science.[56] Today it has migrated into distinct fields, stretching from the humanities to government, and has experienced a renaissance in the contemporary sciences where, as Delfanti has shown, commitments to the ideal are reinvented in response to shifting economic contexts.[57] Like participation, openness tends to be rhetorically invested with such a positive valence that it stands resistant to critique. "Openness is a philosophy that can rationalize its own failure, chalking people's inability to participate up to choice," observes cultural critic Astra Taylor.[58] And many participatory projects claim—loudly and proudly—their openness.

This openness can certainly be a superb mobilizer for producing and sharing knowledge, as Delfanti and Iaconesi's account of the "open source cancer" project, included here, affirms. After being diagnosed with cancer and in an attempt to demedicalize his condition, Iaconesi codevised a website called La Cura with his partner Oriana Persico. He converted all of the medical records related to his brain tumor from proprietary and

professional to user-friendly standards and made them accessible on the website, asking people to reply with "cure" scenarios.
The multiform response—the website received hundreds of
thousands of contributions from physicians, patients, artists,
activists, and others—is but a testimony to the possibility of
using the Internet as a means to sharing information otherwise
inaccessible to laypersons, as well as a means to create rituals of
bio-empowerment. Pullin's essay also shows how it is indeed
possible to codesign augmentative communication devices with
people living with speech disabilities who rely on speech-
generative devices to communicate. "What if speech technology,"
he asks, "were conceived of as an open source medium in a
deeper sense, in which myriad tones of voice are crafted,
exchanged and appropriated by the very people who use it in
their everyday lives?" His essay examines different participatory
projects in augmentative communication, which seek to enrich
expressiveness of speech and tone of voice *with* (as opposed to
for) people with disabilities.

Openness, however, is not a straightforward—much less a
singular—state of being. And success stories like those above
require a high level of expertise in design, programming, or
hacking processes. While numerous participatory media
projects rely on a colloquial understanding of openness—
simply allowing anyone to participate—in practice, openness
is operationalized distinctly as an endeavor. As media scholar
Nathaniel Tkacz has carefully shown, openness can refer to a
procedure internal to a project or it can concern general access
to goods. The problem with a term like openness arises when it
is used alone or in association with concepts that are too closely
related (like transparency and participation). However, by
putting multiple categories into play, such as expertise with
openness, we gain the necessary conceptual traction to more
clearly see power dynamics at work. And indeed, expertise—
commanding particular sets of skills—is necessary for any form
of participation; it limits some and enables others to engage
interdependently, as Christina Dunbar-Hester's essay in this
collection powerfully demonstrates. A cadre of experts—pro-
grammers, designers, system administrators, technically minded

journalists, and policymakers—have risen to become prominent actors in various fields of endeavor. These experts are now important brokers, bridging between existing institutions (such as newspapers or software firms) and newly emergent ones (such as the free software project or citizen-led journalism sites). Thus theories of brokerage and trading zones are also essential for any understanding of contemporary digital participation.[59]

3. *Surveillance.* In the emerging media environment, participation has become an engine of commerce, consent, and control.[60] Contributors, especially those influenced by the work of Michel Foucault, argue that new media participation leads to new forms of cooptation and surveillance by governments, corporations, and other users. Another version of this critique focuses on labor conditions in digital culture, where participation often means that participants provide new media companies with value, either in exchange for "free" entertainment or in exchange for often subminimum wages, as discussed in Scholz's essay in this volume. Andrejevic's essay describes the "passive-ication" of interactivity as the phenomenon whereby communications technologies effectively force people to "participate" in real time with their data—in spite of themselves—while Kate Crawford's essay traces similar dynamics on the scale of urban surveillance systems.

As the essays included in the "Participation under Surveillance" section of this book demonstrate, the stakes of moving away from singular, blanket categories like openness become particularly salient in light of the dominance of corporate platforms like Facebook, whose interest in encouraging sharing, participation, and openness is directly linked to a privacy-violating profit model based on harvesting and reselling personal data for advertising. "If people share more, the world will become more open and connected. And a world that's more open and connected is a better world," Mark Zuckerberg (one of the cofounders of Facebook) famously announced in 2010.[61] Discourses based on sharing and openness in these circumstances occlude—and in a way that uncannily resembles Enlightenment colonialist logics[62]—just how many "acts of communication are now, by definition, acts of surveillance meshed within an economy that aggregates even the affective, non-representational dynamics of relation."[63] It is for this reason that anonymous organizing and piracy, which are proliferating today, have become paramount sites of participatory struggle, where citizens can escape the logic of extraction and surveillance.[64]

Ubiquitous surveillance facilitated by ICTs—what Crawford designates as "algorithmic listening"—and the gathering of personal data currently operated by web-based corporations (commercial surveillance) and governments (the NSA program, for example) are not simply matters of privacy but also of scale and lack of accountability. In her case study of the Boston Calling pilot—a surveillance system used for crowd detection in public spaces following the Boston Marathon bombings in 2013— Crawford discloses that these types of pilots, used to test law enforcement scenarios, "began in off-site locations, military bases, and custom-made environments, but they are now moving to the lived environments of millions of people." More problematically, these scenarios are moving into urban public spaces without the consent of citizens, justified by a rhetoric of "permacrisis." As Andrejevic and Cohen maintain, Internet participation (involving activities such as searching, purchasing, communicating, socially exchanging, or open sourcing) has become a new mode of surveillance beyond any participant's control. This is one of the ultimate paradoxes of what Cohen calls "the participatory turn in surveillance": The more we participate, the more data is gathered about us, and (to paraphrase Andrejevic) the less participatory participation becomes. The participatory condition—in which participation begets surveil-lance—can be compared to philosopher Giorgio Agamben's definition of the apparatus: It constitutes us as subjects, but also robs us of subjectivity in the process. Even if one does not agree with the austerity of this assessment, it is important to point out that Agamben sees *profanation*—"the restitution to common use" or "the free use"—as a means to make apparatuses work correctly.[65]

4. *Aisthesis.* In the aesthetic regime, the irruption of equality collapses hierarchies of genres and styles. As Jean-Philippe

Deranty has pointed out in his analysis of Rancière's argument, equality not only "reshapes the very modes of perception and thought" of the previous representative regime of art, it more importantly "opens the entire field of *aisthesis,* the world itself as something to be sensed, perceived and thought, for modes of expression to be reinvented."[66] The essays in this volume that investigate the aisthesis (αἴσθησις, the faculty to perceive by the senses, as well as by the intellect) generated in participatory media art rekindle debates around the aesthetic regime, especially its manifold contestations of the separation of art and life. They also pose more pragmatic questions, relevant to participation in other spheres, regarding who can take part in the creative process, what constitutes the nature of a creative collaboration, and what specific forms these practices generate.[67] These essays all insist on the need to revisit participatory practices through the reinvention of dialogue—interperception, transmission, and storytelling. The main argument here is not that participatory processes in the field of art constitute a form of political action that can change the world, but that they might in certain circumstances generate new perceptions of, in, and about the world—since an element of contingency always persists in any situation where human agency is at play.

For his major curatorial retrospective on "The Art of Participation," Rudolf Frieling notes that while the participatory artwork "requires your input and your contribution," "you watch others and others watch you."[68] The spectacle of participation becomes intrinsic to the work. In his essay for this volume, he turns his attention to a recent work by Dora García to show how collaborative art can be productively refashioned when watching others and being watched by others becomes the very subject of the work. How are such environments aesthetically innovative? And how can a perceptual dialogue between participants be inventive and nonformulaic? The revisiting of dialogue as a form of conversation is common to all of the essays concerned with aisthesis. Jason E. Lewis's essay describes the activities of the Aboriginal Territories in Cyberspace (AbTec) Lab he codirects with Skawennati. AbTec seeks to integrate traditional indigenous storytelling to new media sites and video games. Working in collaboration with a younger generation of indigenous storytellers, Lewis and Skawennati have produced narratives that counter the phantasm of the "imaginary Indian" to build what they call a "future imaginary." Bernard Stiegler's essay also centers on the question of transmitting knowledge across generations. He argues that universities must learn to use new media in order to ensure a form of transmission that does not simply consist in the reproduction of knowledge, but that generates an *anamnésis* (a transformative reminiscence of knowledge) through transindividuation—the transmission of knowledge through dialogue and debate between protagonists who have learned to think by themselves and to deliberate accordingly.

We have chosen to close this section, as well as this volume, with a portfolio of a work by the artists Rafael Lozano-Hemmer and Krzysztof Wodiczko. *Zoom Pavilion* (2013), an augmented reality interactive installation initially conceived for the Fifth China International Architectural Biennial, relies on the assumption that participation becomes successful only when it is "out of control."[69] As people walk into the illuminated public space of *Zoom Pavilion,* they are detected by computerized tracking systems that establish their position, velocity, and acceleration. Their image is immediately projected on the ground next to them at a normal scale of 1:1, but then amplified (with up to 35× magnification) by robotic cameras as they zoom in. Making full use of surveillance mechanisms, the installation operates an aesthetic *détournement* of that technology, disorienting the public's relation to its own image. The public space that is envisioned in this work is one in which human participants coexist with mutating abstractions of themselves that they must learn to converse with. To this date, their project remains unrealized.

Participation in the Age of Consensus

It might well be that participatory practices have generated a condition not only because of their expansion throughout societal fields, but more decisively because participation today hails individuals and publics as subjects in the social order of *consensus.* This order represents the waning of the political as an activity of the possible—a process that has been evolving at a global scale since the collapse of the Berlin Wall and the related dissolution of communism as one of the last political narratives of emancipation. In line with what Erik Swyngedouw has said about sustainability, participation—whether taken as a concept or as a practice—is now so bereft of political content and so elevated as a moral value that it is impossible to disagree with its formulation, goals, and promises of a better life.[70]

One thing that seems clear is that established holders of economic and political power (i.e., capitalist corporations and the state) have adapted very well to the participatory condition and actually thrive under it. In the emerging media environment, participation has become a preferred engine of commerce, consent, and control.[71] The cunning of participation is that it seldom feels like any of these, because it is what we, as free liberal individuals and self-governing democratic citizens, have come to expect. As Andrew Barry has observed, participation—here styled as "interactivity"—is a technique that aligns perfectly with the rise of neoliberal practices. Individuals are interpellated as self-regulating subjects who don't just participate in politics but, rather, govern themselves by participation. In this case, participation ceases to be a check on political power and instead becomes a model for its exercise. "Active, responsible and informed citizens have to be made," Barry writes. "Today, interactivity has come to be a dominant model of how objects can be used to produce subjects. In an interactive model, subjects are not disciplined, they are allowed."[72] The question is: allowed to do what? For it is not at all clear that being allowed to participate amounts to being allowed to appear as one wishes to appear, to have an equal share, to think, to disagree fundamentally, to oppose, to abstain, to dissent, to deliberate, to judge, to decide, to organize, to act, to create something new, or to do any of the other things we might suppose a political being ought to be able to do. If intellectuals in the 1950s challenged the ways media institutions invited audiences to watch, to listen, and to engage through consumption, today the tables have turned. We must challenge media institutions' constant demands to interact and to participate, as if those activities were seen as fulsome by dint of their very nature.

This suggests the deep political ambiguity of the participatory condition. It is always disorienting when something you thought you loved becomes loved by those whom you do not love so much. We have loved participation for a very long time, and have fought fiercely to gain and secure it. Now we (or, at least, some of us) have managed to attain it.

And just as we have begun to exercise it intensively and ubiquitously, it turns out that others love it too: bureaucracies, police forces, security and intelligence agencies, and global commercial enterprises, among others. This is the political agony of the participatory condition: It can be neither embraced nor disavowed without considerable loss. We are not happy with participation, but were we to lose it, we would be sad. It is thus the name of our collective melancholy, a condition marked by what Wendy Brown described, in reference to liberal democracy more generally, as "a dependency we are not altogether happy about, an organization of desire we wish were otherwise."[73] It might be best to begin the hard work of discerning and materializing that otherwise, but we are not yet in a position to do so. For now, there is Gayatri Spivak's suggestion that such situations demand "a persistent critique of what we cannot not want."[74] Such a critique is necessary, but it is hardly satisfying. It reflects the impasse to which the participatory condition brings us: In the prevailing language and practice of our democratic convictions and aspirations, "participation" becomes a security against the possibility of their substantive realization. Under the participatory condition, democratic politics turns against itself, fulfilling the diagnosis made by Rancière in *On the Shores of Politics:* "Depoliticization is the oldest task of politics, the one which achieves its fulfillment at the brink of its end, its perfection on the brink of the abyss."[75] What the participatory condition finally demands of us is that we struggle to think and act our way beyond this abyss.

Notes

1. Definitions of the term "participation" in Indo-European languages are generally anchored in the idea of shared action, invoking notions of *taking part* or *effecting an action with* something or someone else. For instance, the French definition of the verb *participer* reads as *prendre ou avoir part à,* and finds a direct equivalent in the German verb *teilnehmen,* which literally means to take *(nehmen)* part *(teil),* while the verb's synonym, *mitmachen,* transliterates in English as "doing with." Like the Latin *participare,* derived from a combination of words declined from *pars* (part) and *capere* (to take), the Greek συμμετέχω combines the prefix σύν (with) and the verb μετέχω (to take part).

2. Henry Jenkins, *Textual Poachers: Television Fans and Participatory Culture* (New York: Routledge, 1992).

3. Henry Jenkins, Sam Ford, and Joshua Green, *Spreadable Media: Creating Value and Meaning in a Networked Culture* (New York: New York University, 2013), 7.

4. Pierre Rosanvallon, *The Society of Equals*, trans. Arthur Goldhammer (Cambridge, Mass.: Harvard University Press, 2013), 10.

5. Louis Althusser, "Ideology and Ideological State Apparatuses (Notes towards an Investigation)," in *Lenin and Philosophy and Other Essays*, trans. Ben Brewster (New York: Monthly Review Press, 1972), 174.

6. Henry Jenkins with Ravi Purushotma, Margaret Weigel, Katie Clinton, and Alice J. Robison, eds., *Confronting the Challenges of Participatory Culture: Media Education for the 21st Century* (Cambridge, Mass.: MIT Press, 2009), 7.

7. Ernest Barker, ed. and trans., *The Politics of Aristotle* (Oxford: Oxford University Press, 1962), 93.

8. On the preference not to participate, see Herman Melville's iconic "Bartleby, The Scrivener," in *The Piazza Tales* (New York: Dix and Edwards, 1856). For a critical treatment of the political implications of Bartleby's apparent passivity, see Giorgio Agamben, *Homo Sacer: Sovereign Power and Bare Life* (Stanford, Calif.: Stanford University, 1998). See also Nathalie Casemajor, Stéphane Couture, Mauricio Delfin, Matt Goerzen, and Alessandro Delfanti, "Non-Participation in Digital Media. Toward a Framework of Mediated Political Action," in *Media, Culture & Society* 37, no. 6 (2015): 850–66.

9. Jacques Rancière, *Disagreement: Politics and Philosophy* (Minneapolis: University of Minnesota Press, 1999), 11.

10. See Nico Carpentier, *Media and Participation: A Site of Ideological-Democratic Struggle* (Bristol: Intellect, 2011), 15–38.

11. Hannah Arendt, *The Human Condition* (Chicago: University of Chicago Press, 1958).

12. Jürgen Habermas, *The Structural Transformation of the Public Sphere: An Inquiry into a Category of Bourgeois Society,* trans. Thomas Burger (Cambridge, Mass.: MIT Press, 1991).

13. See Matt Ratto and Megan Boler, eds., *DIY Citizenship: Critical Making and Social Media* (Cambridge, Mass.: MIT Press, 2014).

14. C. B. Macpherson, *The Life and Times of Liberal Democracy* (Oxford: Oxford University Press, 1977); Carole Pateman, *Participation and Democratic Theory* (Cambridge: Cambridge University Press, 1970); Ernesto Laclau and Chantal Mouffe, *Hegemony and Socialist Strategy: Toward a Radical Democratic Politics* (London: Verso, 1985); Benjamin Barber, *Strong Democracy: Participatory Politics for a New Age* (Berkeley: University of California Press, 1984).

15. David Graeber, *The Democracy Project: A History, a Crisis, a Movement* (London: Penguin, 2013).

16. Jenkins et al., *Spreadable Media*, 160–61, 193.

17. C. Wright Mills, *The Power Elite* (Oxford: Oxford University Press, 1956). Herbert Marcuse, *One-Dimensional Man* (Boston: Beacon Press, 1964). Guy Debord, *The Society of the Spectacle,* trans. Ken Knabb (New York: Rebel, 1983).

18. Christopher Lasch, *The Culture of Narcissism: American Life in an Age of Diminishing Expectations* (New York: W. W. Norton, 1991); Robert Bellah, et al., *Habits of the Heart: Individualism and Commitment in American Life* (Berkeley: University of California Press, 1985).

19. Michael Sandel, *Liberalism and the Limits of Justice* (Cambridge: Cambridge University Press, 1981); Charles Taylor, *Sources of the Self: The Making of the Modern Identity* (Cambridge: Cambridge University Press, 1989); Michael Walzer, *Spheres of Justice* (Oxford: Blackwell Press, 1983); Robert Putnam, *Bowling Alone: The Collapse and Revival of American Community* (New York: Simon & Schuster, 2000).

20. Nikolas Rose, *Powers of Freedom: Reframing Political Thought* (Cambridge: Cambridge University Press, 1999).

21. Darin Barney, "Publics without Politics: Surplus Publicity as Depoliticization," in *Publicity and the Canadian State: Critical Communications Approaches,* ed. Kirsten Kozolanka (Toronto: University of Toronto Press, 2013), 70–86.

22. Jacques Rancière, *Aesthetics and Its Discontents,* trans. Steven Corcoran (Cambridge: Polity Press, 2009), footnote 6, 11.

23. UN General Assembly, *Universal Declaration of Human Rights,* December 10, 1948, 217A (III): Article 27.

24. For instance, all of the live arts traditions, from music to theater and dance, have developed their own sets of conventions for group improvisation, essentially establishing guidelines for participation in a common, though as of yet undefined, aesthetic project. Cf. Keith Johnstone, *IMPRO: Improvisation and the Theatre* (London: Faber and Faber, 1979); Ann Cooper Albright and David Gere, eds., *Taken by Surprise: A Dance Improvisation Reader* (Middletown, Conn.: Wesleyan University Press, 2003).

25. For Claire Bishop, the idea that "art should be useful and effect concrete changes in society" (see: *Artificial Hells: Participatory Art and the Politics of Spectatorship* [London: Verso, 2012], 52) finds its source in the Russian Proletkult theater, as well as in Futurist *seratas* and in the performative excursions and trials of the Paris Dada (see chapter 2 in *Artificial Hells*). Grant Kester also acknowledges the relevant "legacy of modernist art" in examples of contemporary dialogical art, such as WochenKlausur's *Intervention to Aid Drug-Addicted Women.* For Kester, the "relevant legacy of modernist art" is to be found "not in its concern with the formal conditions of the object, but rather in the ways in which aesthetic experience can challenge conventional perspectives . . . and systems of knowledge." Grant Kester, *Conversation Pieces: Community and Conversation in Modern Art* (Berkeley: University of California Press, 2004), 3.

26. Brecht's *Verfremdungseffekt* relies on specific aesthetic strategies: dropping the convention of the theatrical "fourth wall"; using the principle of montage as a structuring element, and including frequent interruptions as well as contradictory elements within the primary narrative; asking actors to rehearse their dialogues in the third person or in the past tense in order to reinforce their distance from the characters portrayed; having the stage directions spoken aloud; using placards to announce the action in scenes to come; and further devices, such as the use of nonillusionistic sets and the inclusion of music so as to disrupt any sense of naturalism that might be conveyed onstage. See Bertolt Brecht and John Willett, *Brecht on Theatre: The Development of an Aesthetic* (New York: Hill and Wang, 1964).

27. In Boal's conception of the theatre, "there are no *spectators,* only *active observers* (or spect-actors). The center of gravity is in the auditorium, not on the stage." The aims of the techniques developed for his "Theater of the Oppressed" are twofold: "(a) to help the spect-actor transform himself into a protagonist of the dramatic action and rehearse alternatives for his situation, so that he may then be able (b) to extrapolate into his real life the actions he has rehearsed in the practice of theatre." Augusto Boal, *The Rainbow of Desire* (London: Routledge, 1995), 40.

28. Term introduced by Hans-Thies Lehmann in 1999 to describe "the profoundly changed mode of theatrical sign usage" since the 1960s, referring in particular to the rise of new theater forms that rely on neither plot nor character. The characteristics of postdramatic theater overlap with some of the conventions found in performance art and experimental theater, as well as physical theater and dance. It is suggested that, as compared to narrative-driven theater, in postdramatic theater a greater demand is placed on the audience to participate in the meaning-making process. Hans-Thies Lehmann, *Postdramatic Theatre* (London: Routledge, 2006), 17.

29. Some of the most prominent terms for these practices include: *arte util* or "useful art" (Tania Brughera); collaborative art; community art; conversational art (Homi Bhabha); cultural activism; dialogical art (Grant H. Kester); do-it-yourself art (Anna Dezeuze); *médiation culturelle;* new audience development; new genre public art (Suzanne Lacy); placemaking (Ronald Lee Fleming); public practice; relational aesthetics (Bourriaud); socially engaged art; social practice; and social sculpture (Joseph Beuys). See also Nico Carpentier, "The Arts, Museums and Participation," in *Media and Participation: A Site of Ideological-Democratic Struggle* (Bristol: Intellect, 2011), 55–64.

30. Anna Dezeuze, ed., *The "Do-It-Yourself" Artwork: Participation from Fluxus to New Media* (Manchester: Manchester University Press, 2010), 15.

31. Beryl Graham, "What Kind of Participative System? Critical Vocabularies from New Media Art," in *The "Do-It-Yourself" Artwork*, ed. Anna Dezeuze, 281–305.

32. Bishop, *Artificial Hells*, 8.

33. Jennifer Barrett, *Museums and the Public Sphere* (Oxford: Blackwell, 2012); and Nina Simon, *The Participatory Museum* (Santa Cruz, Calif.: Museum 2.0, 2010).

34. Though the idea of community remains ambiguous in the context of "community-based collaborations," Miwon Kwon observes that it generally falls within one of the following four typologies: "community of mythic unity; 'sited' communities; temporary invented communities; and ongoing invented communities." Miwon Kwon, *One Place after Another: Site-Specific Art and Locational Identity* (Cambridge, Mass.: MIT Press, 2002), 6–7.

35. Grant H. Kester, *Conversation Pieces: Community and Communication in Modern Art* (Berkeley: University of California Press, 2004), 29.

36. Ibid., 29–30.

37. This is the critique Tony Fisher puts forward regarding Augusto Boal's conception of "metaxis." Tony Fisher, "The Arraignment of Power: Augusto Boal and the Emergence of the Radical Democratic Theatre Subject," *Performance Research* 16, no. 4 (2001): 20.

38. John Durham Peters, *Speaking into the Air: A History of the Idea of Communication* (Chicago: University of Chicago Press, 1999); Habermas, *Structural Transformation of the Public Sphere;* Richard John, *Network Nation: Inventing American Telecommunications* (Cambridge, Mass.: Harvard University Press, 2010).

39. Michael Denning, *Mechanic Accents: Dime Novels and Working Class Culture in America* (New York: Verso, 1983); Peter Hinks, *To Awaken My Afflicted Brethren: David Walker and the Problem of Antebellum Slave Resistance* (University Park: Pennsylvania State University Press, 1997).

40. Michèle Martin, *"Hello, Central?": Gender, Technology and Culture in the Formation of Telephone Systems* (Montreal: McGill-Queens University Press, 1991); Carolyn Marvin, *When Old Technologies Were New: Thinking about Electrical Communication in the Nineteenth Century* (New York: Oxford University Press, 1988); Mary Ryan, *Women in Public: Between Banners and Ballots, 1825–1880* (Baltimore: Johns Hopkins University Press, 1990); Greg Downey, *Telegraph Messenger Boys: Labor, Technology and Geography, 1850–1950* (New York: Routledge, 2002).

41. George Simmel, *The Philosophy of Money,* trans. Tom Bottomore and David Frisby (New York: Routledge, 1990).

42. Friedrich Kittler, *Gramophone-Film-Typewriter,* trans. Geoffrey Winthrop-Young and Michael Wutz (Stanford, Calif.: Stanford University Press, 1999), 57; George Simmel, *The Philosophy of Money,* 257.

43. Siegfried Kracauer, *The Mass Ornament: Weimar Essays,* trans. Thomas Y. Levin (Cambridge, Mass.: Harvard University Press, 2005); Theodor Adorno, "The Radio Symphony: An Experiment in Theory," in *Radio Research 1931,* eds. Paul Lazarsfeld and Frank Stanton (New York: Columbia Office of Radio Research, 1941), 110–39; Marshall McLuhan, *The Mechanical Bride: Folklore of Industrial Man* (New York: Vanguard Press, 1951); Gunther Anders, "The Phantom World of TV,"

in *Mass Culture: The Popular Arts in America,* eds. Bernard Rosenberg and David Manning White (New York: Free Press, 1957), 358–67; Georg Lukacs, *History and Class Consciousness* (Cambridge, Mass.: MIT Press, 1971); Guy Debord, *Society of the Spectacle,* trans. Ken Knabb (New York: Rebel, 1983).

44. Charles Horton Cooley, *Social Organization: A Study of the Larger Mind* (Glencoe: Free Press, 1909), 80.

45. Ibid., 87–88.

46. Frantz Fanon, *A Dying Colonialism,* trans. Haakon Chevalier (New York: Monthly Review Press, 1965); Paddy Scannell, "For Anyone-as-Someone Structures," *Media, Culture and Society* 22, no. 1 (2000): 5–24; Michelle Hilmes, *Radio Voices: American Broadcasting 1922–1952* (Minneapolis: University of Minnesota Press, 1997).

47. Walter Lippmann, *The Phantom Public* (New York: Harcourt, Brace, 1925); Daniel Goldhagen, *Hitler's Willing Executioners: Ordinary Germans and the Holocaust* (New York: Verso, 1996); Lisa Parks, "Insecure Airwaves: US Bombings of Al-Jazeera," *Journal of Communication and Critical/Cultural Studies* 4, no. 2 (2007): 226–31.

48. Tim Ingold, *Being Alive: Essays on Movement, Knowledge and Description* (London: Routledge, 2011), 30.

49. Clay Shirky, *Here Comes Everybody* (New York: Penguin Books, 2009). See also Yochai Benkler, *The Wealth of Markets: How Social Production Transforms Markets and Freedom* (New Haven, Conn.: Yale University Press, 2006); Eric von Hippel, *Democratizing Innovation* (Cambridge, Mass.: MIT Press, 2005); Alex Bruns, *Blogs, Wikipedia, Second Life, and Beyond: From Production to Produsage* (New York: Peter Lang, 2008).

50. Max Weber, *Economy and Society* (Berkeley: University of California Press, 1978).

51. Adam Fish, Luis F. R. Murillo, Lilly Nguyen, Aaron Panofsky, and Christopher M. Kelty, "Birds of the Internet: Towards a Field Guide to the Organization and Governance of Participation," *Journal of Cultural Economy* 4, no. 2 (2011): 157–87.

52. Nico Carpentier, *Media and Participation: A Site of Ideological-Democratic Struggle* (Bristol: Intellect, 2011).

53. Howard Rheingold, *The Virtual Community* (New York: HarperCollins, 1993); Steve Jones, ed., *Cybersociety 2.0: Revisiting CMC and Community* (Newbury Park, Calif.: Sage, 1998); Richard Barbrook, "The High-Tech Gift Economy," *First Monday* Special Issue #3 (December 2005), http://firstmonday.org/ojs/index .php/fm/article/view/1517/1432/; Sadie Plant, *Zeroes and Ones: Digital Women and the New Technoculture* (New York: Fourth Estate Limited, 1998).

54. Henry Jenkins, *Convergence Culture: Where Old and New Media Collide* (New York: New York University Press, 2006); Chris Kelty, *Two Bits* (Durham, N.C.:

Duke University Press, 2008); Gabriella Coleman, *Coding Freedom: The Ethics and Aesthetics of Hacking* (Princeton, N.J.: Princeton University Press, 2012); Zeynep Tufekci and Christopher Wilson, "Social Media and the Decision to Participate in Political Protest: Observations from Tahrir Square," *Journal of Communication* 62 (2012): 363–79. Cathy Davidson, *Now You See It: How Technology and Brain Science Will Transform Schools and Business for the 21st Century* (New York: Penguin Books, 2012).

55. Nathaniel Tkacz, "From Open Source to Open Government: A Critique of Open Politics," *Ephemera: Theory and Politics in Organization* 12, no. 4 (2012): 386–40.

56. William Eamon, "From the Secrets of Nature to Public Knowledge: The Origins of the Concept of Openness in Science," *Minerva* 23, no. 3 (1985): 321–47.

57. Alessandro Delfanti, *Biohackers: The Politics of Open Science* (London: Pluto Press, 2013).

58. Astra Taylor, *The People's Platform: Taking Back Power and Culture in the Digital Age* (New York: Metropolitan Books, Henry Holt and Company, 2014).

59. Peter Galison, *Image & Logic: A Material Culture of Microphysics* (Chicago: University of Chicago Press, 1997).

60. Jodi Dean, *Democracy and Other Neoliberal Fantasies* (Durham, N.C.: Duke University Press, 2009).

61. Mark Zuckerberg, "From Facebook, Answering Privacy Concerns with New Settings," *The Washington Post,* May 24, 2010, http://www.washingtonpost .com/wp-dyn/content/article/2010/05/23/AR2010052303828.html/.

62. Anita Chan, *Networking Peripheries: Technological Futures and the Myth of Digital Universalism* (Cambridge, Mass.: MIT Press, 2014).

63. Ned Rossiter and Soenke Zehl, "Privacy Is Theft: On Anonymous Experiences, Infastructural Politics, and Accidental Encounters," http://nedrossiter.org /?p=374/.

64. Ibid.

65. Giorgio Agamben, *What Is an Apparatus? And Other Essays,* trans. David Kishik and Stefan Padetella (Stanford, Calif.: Stanford University Press, 2007), 14, 24, and 28.

66. Jean-Philippe Deranty, "The Symbolic and the Material: A Review of Jacques Rancière's *Aisthesis: Scenes from the Aesthetic Regime of Art* (Verso 2013)," *Parrhesia* (18), 2013, http://www.parrhesiajournal.org/parrhesia18/parrhesia18 _deranty.pdf/.

67. Theodor W. Adorno, Gretel Adorno, and Rolf Tiedemann, *Aesthetic Theory* (Minneapolis: University of Minnesota Press, 1997).

68. Suzanne Stein, "Interview: Rudolf Frieling on the Art of Participation," November 5, 2008, http://openspace.sfmoma.org/2008/11/interview-rudolf -frieling-on-the-art-of-participation/.

69. Rafael Lozano-Hemmer, "Antimonuments and Subsculptures," keynote presentation at *The Participatory Condition* colloquium, November 14, 2013.

70. Erik Swyngedouw, R. Krueger, and E. Gibbs, eds., *The Sustainable Development Paradox: Urban Political Economy in the United States and Europe* (London: Guildford Press, 2007), 13–40.

71. Dean, *Democracy and Other Neoliberal Fantasies.*

72. Andrew Barry, *Political Machines* (London: Athlone, 2001), 129; Quentin Skinner, *The Foundations of Modern Political Thought*, Vol. 1: *The Renaissance* (Cambridge: Cambridge University Press, 1978).

73. Wendy Brown, "Neo-Liberalism and the End of Liberal Democracy," *Theory and Event* 7, no. 1 (2003): 13.

74. Gayatri Chakravorty Spivak, *Critique of Postcolonial Reason: Toward a History of the Vanishing Present* (Cambridge, Mass.: Harvard University Press, 1999), 110.

75. Jacques Rancière, *On the Shores of Politics* (London: Verso, 1995), 19.

· I ·

The Politics of Participation

CHAPTER 1

Power as Participation's Master Signifier

Nico Carpentier

THIS CHAPTER SETS OUT to argue that participation is a political-ideological concept that is intrinsically and intimately linked to power. This close connection is symbolized by reverting to Lacan's metaphor of the master signifier—without blindly respecting Lacanian orthodoxy—in order to argue that the notion of power stabilizes (or allows the stabilization of) the incessant contingency of "participation's" meaning. The wide variety of meanings attributable to participation, or its conceptual instability, already featured prominently on the opening page of Pateman's 1970 book *Democratic Theory and Participation,* when she wrote: "the widespread use of the term . . . has tended to mean that any precise, meaningful content has almost disappeared."[1]

In analyzing the ways these conceptual contingencies of participation are handled—including Pateman's strategy to (re)introduce precise meanings for participation—this chapter distinguishes between two main strategies: a dichotomizing strategy that aims to distinguish "real" participation from "fake" or "pseudo" forms of participation, and a strategy that embraces the radical contingency of the concept and articulates it as a site of semantic-democratic struggle between more minimalist and more maximalist versions.[2] Despite the differences, both strategies share a strong focus on power as participation's defining characteristic, which allows me to define participation as a (formal or informal) decision-making process which involves nonprivileged and privileged actors whose power relationships are (to some extent) egalitarian.

Despite the similarities between the two strategies, and without discrediting the first strategy, this chapter will align itself with the second strategy and its focus on participation as a site of semantic-democratic struggle. The presence of this struggle will be illustrated first by a discussion of democratic theory, where participation is in permanent tension with the concept of representation. A relatively small but important detour into feminism and participation will then show that the articulation of participation as a site of struggle is not limited to the field of institutionalized politics, but is being played out in a variety of societal spheres. This will pave the way for a more extensive discussion of media participation, where we again can find a struggle between more minimalist and more maximalist participatory models. What this chapter ultimately aims to show is that the political nature of participation manifests itself in the discursive and material struggles (and their combinations) to minimize or to maximize the equal power positions of the actors involved in decision-making processes, and that these power struggles over power itself are omnipresent in all societal spheres.

Coping Strategy One

The complexity and contingency of participation requires theoretical frameworks to deploy coping mechanisms, and I would argue here that there are two main ways of dealing with this contingency and complexity. The first strategy is based on an expression of regret regarding participation's significatory chaos, combined with the attempt to undo this chaos through an (almost archaeological) unraveling of participation's authentic meaning. This strategy leads to the construction of dichotomized systems of meaning, in which specific forms of participation are described as "real" and "authentic," while other forms are described as "fake" and "pseudo." This strategy is relatively old; in the field of political participation, for example, Verba pointed to the existence of pseudoparticipation, in which the emphasis is not on creating a situation in which participation is possible, but on creating the feeling that participation is possible.[3] An alternative label, used by Strauss among others, is manipulative participation.[4]

Other, but still similar, strategies arose out of the construction of hierarchically ordered and multilayered systems, of which Arnstein's ladder of participation is a prime example.[5] In her seminal article "A Ladder of Citizen Participation," she linked participation explicitly to power, writing "that citizen participation is a categorical term for citizen power."[6] She continues:

> It is the redistribution of power that enables the have-not citizens, presently excluded from the political and economic processes, to be

deliberately included in the future. It is the strategy by which the have-nots join in determining how information is shared, goals and policies are set, tax resources are allocated, programs are operated, and benefits like contracts and patronage are parceled out.[7]

Arnstein developed a categorization of participation (the "ladder"), in which she distinguished three main categories (citizen power, tokenism, and nonparticipation) and eight levels. The category of nonparticipation consists of two levels: manipulation and therapy. Tokenism has three levels: informing, consultation, and placation. The last category is citizen power, which has three levels: partnership, delegated power, and citizen control. This categorization system allows her to distinguish between processes that are participatory and processes that are not (tokenism and nonparticipation).

Another classic definition of participation that used this dichotomizing strategy was developed by an author who has been mentioned before, namely Pateman in her 1970 book *Democratic Theory and Participation.* Pateman distinguished between partial and full participation, where partial participation is defined as "a process in which two or more parties influence each other in the making of decisions but the final power to decide rests with one party only."[8] Full participation is seen as "a process where each individual member of a decision-making body has equal power to determine the outcome of decisions."[9] These definitions and approaches share an almost messianic concern for the concept of participation: They want to protect and rescue it. The tactics are relatively similar, because they all consist of differentiating between ("authentic" or "real") participation and other practices that are only nominally participatory—and can be unmasked as forms of pseudoparticipation.

There are problems with this first strategy, which can be symbolized by Jan Fabre's artwork *The Man Who Measures the Clouds.* Positioned on the rooftop of several buildings in Belgium, the United States, and Japan, the statue portrays a man standing on a small ladder, reaching out to the sky with a large ruler in both hands, attempting to measure the clouds, and permanently failing to do so. Quite often, these ladder-based approaches suggest the existence of easy cutoff points between dichotomized positions. Even when several steps or levels are distinguished, these discrete models still suggest fairly crude categorizations which do not always rest well with the complexities of participatory practices. Second, the multilayeredness of participatory processes also makes them difficult to capture by the ladder-based approaches. Participatory intensities can change over time, but several components within one process can sometimes also yield differences. In his discussion of participatory (open) ethics, Ward explains how participation in a specific process might be intense in one component, but minimal in another.[10] For instance, participatory (open) ethics could be open in the *discussion* of new ethical guidelines, but not in their formal adoption. Often, Ward argues, we can "only reach a rough, comparative judgment," especially when "there are forces pulling in opposite directions."[11] Take, for instance, YouTube: The site allows for participation in publishing videos, but not in the management of YouTube itself. To quote Jenkins on this matter: There are "limits to our ability to participate in You-Tube—the degree to which participants lack any direct say in the platform's governance. This is very different from discussing how participatory communities might use YouTube as a distribution channel."[12]

The multilayeredness of participation becomes even clearer when a particular process in one field facilitates participation in another. Here we can build on Wasko and Mosco's distinction between democratization *in* and *through* the media, so that we can distinguish between participation *in* a particular field and participation *through* a particular field.[13] For instance, reader participation in the newspaper organization might be very rare (with some notable exceptions), but newspapers allow readers to intervene (to some degree) in the political field. Likewise, participation in a particular artistic process might again be limited, but through this process visitors' participation in (and thus control over) the urban setting might be facilitated. There are—what I would propose to call—*transgressed* forms of participation (where the participatory process transgresses the boundaries of a particular field and becomes situated in several fields) and *transferred* forms of participation (where a nonparticipatory process in a particular field allows for participation in another field) that are difficult to capture by ladder-based approaches. Third, the ladder-based approaches tend to see participation as the stable outcome of a process, ignoring the struggles over participatory intensities within these processes, within particular fields, and within society. Different actors might have different perspectives and interests, and will develop different strategies to see their perspectives realized, thus entering into conflict with each other. Arguably, this generates a much more dynamic and contingent (or unstable) process than the ladder-based approaches seem to suggest.

Not Throwing Out the Baby with the Bathwater

Despite these problems, the first approach remains valuable. More specifically, three components of the first strategy, which have to do with three different delineations, are worth salvaging. A first delineation is related to what can be seen as the core issue in this debate on participation (and the political-ideological struggle that lies behind the debate): power. Power—and, more specifically, the way power is distributed in society—is omnipresent in all debates on participation (see Carpentier's *Media and Participation* for a detailed analysis), and is also a constituting element of the first strategy. I will return to this in the next part of this chapter.

The second delineation that can be derived from the first strategy is related to the difference between participation and its conditions of possibility. Although it remains crucial not to ignore the contingency and structural openness of the participation signifier, some form of discursive fixity is required in order to allow for this concept to be analyzed (and used). This analytical problem can be remedied by again returning to the first approach and to the idea of the prerequisite or the condition of possibility. A number of authors have discussed these so-called prerequisites of participation. Dahlgren mentions civic cultures; Carpentier, Schrøder, and Hallett refer to trust and literacy; and in *Media and Participation*, access and interaction are defined as prerequisites of participation.[14]

Arguably, notions such as access, interaction, and participation are still very different—both in their theoretical origins and in their respective meanings. But they are often integrated (or conflated) into definitions of participation. One example here is Melucci's definition; he says that participation has a double meaning: "It means both taking part, that is, acting so as to promote the interests and the needs of an actor as well as belonging to a system, identifying with the 'general interests' of the community."[15] However valuable these approaches are, I would like to argue that participation is structurally different from its prerequisites—particularly from access and interaction. A negative-relationist strategy—which means focusing on the differences between these three concepts—can then be used to clarify the meaning(s) of participation and to prevent the link with the main defining component of participation, namely power, from being obscured. Moreover, conflating these concepts often causes the more maximalist meanings of participation (see below) to remain hidden, an eventuality which I also want to avoid. From this perspective, the conflation of access, interaction, and participation is actually part of the struggle between the (very) minimalist and maximalist articulations of participation, which again complicates the exact delineation between access and interaction on the one hand, and participation on the other.

The third delineation refers to the emancipatory nature of many of these projects, in their attempt to defend a more maximalist version of participation, and to critique the use of a mere rhetoric of participation. Although I also would like to argue that the minimalist/maximalist dimension has analytical capabilities that can be used by authors grounded in a variety of ideological positions, at the same time I subscribe to the call of many authors (Giddens being one)[16] to continue deepening democracy and including all societal fields in this democratization process. This implies also that I consider the maximalist versions of participation to be socially beneficial attempts to improve the democratic quality of the social, without remaining blind to their problems and challenges. I realize that this plea for an increase in societal power balances has a clear utopian dimension. Situations of full participation, as described by Pateman, are utopian nonplaces (or better, "never-to-be places") which will always be unattainable and empty, but which simultaneously continue to play a key role as ultimate anchoring points and horizons for my work (and the work of many others). Despite the impossibility of fully realizing these situations in social praxis, their fantasmatic realization serves as breeding ground for democratic renewal. But my own normative positioning should not prevent us from seeing the unavoidability of the ideological positioning of any author who intervenes in these debates, whether she defends (or normalizes) a minimalist or a maximalist position. Ideology does not stop at the edges of analyses; it is an integrated part of any analysis, one which requires that authors clearly and explicitly position themselves.

The Focus on Power

As the above-mentioned discussion of the ladder-based approaches has shown, to discuss participation implies talking about actors and their power positions toward each other, and toward production and distribution infrastructures and technologies, within the context of their specific social, political, historical, and economic conditions. This argument unavoidably brings us to the issue of power, and the complexities inherent in this concept. One way of approaching this complexity is through the work of Foucault, who during the 1970s rejected an exclusively restrictive meaning of power, and defined power as productive—as "a general matrix of force relations at a given time, in a given society."[17] Foucault focused on power relations, which he saw as mobile and multidirectional: Power is practiced and not possessed.[18] At the same time, Foucault explicitly stressed that this does not imply that power relations are necessarily egalitarian; rather, the point is that domination alone should not be simplistically considered the "essence" of power.[19] As no actor, however privileged, can exercise full and total control over the social—and more dominant positions will often generate resistance—the Foucauldian model presents us with a multitude of strategies that together form a complex power game, which features contestation, negotiation, and struggle.

Through the work of Laclau and Mouffe we can put (even) more emphasis on the *discursive* struggles that originate in the continuous interplay of power strategies.[20] This antagonistic dimension of the social, or its "context of conflictuality,"[21] is characteristic of the political, and as such a "dimension that is inherent to every human society and that determines our very ontological condition."[22] This position is hardly surprising, given Laclau and Mouffe's poststructuralist emphasis on contingency and difference, which starts from the idea that rearticulations and reconfigurations of the social are always possible. The social is structurally contingent and susceptible to change, but at the same time Laclau and Mouffe recognize that stability and fixity do exist—albeit not without qualification; Laclau and Mouffe emphasize that these fixations are not natural, but result from political interventions. When a particular order becomes strongly fixated, thus benefiting from the luxury of normalization and taken-for-grantedness, Laclau and Mouffe use the concept of hegemony to describe this order. Hegemony is never total, though; as Mouffe writes: "Every hegemonic order is susceptible of being challenged by counterhegemonic practices, i.e., practices which will attempt to disarticulate the existing order so as to install other forms of hegemony."[23] What this approach shows is that participation can be seen as a floating signifier in political discourse, which emphasizes the contingent nature of participation and renders it part of a discursive struggle.

But we should not ignore the fact that this struggle also has material dimensions. This is not in contrast with Laclau and Mouffe's position, as they emphasize the *radical material* component of discourse.[24] For this idea we can find further support in the work of critical discourse analysts

Norman Fairclough and Ruth Wodak. They formulate these relationships as follows:

> Describing discourse as social practice implies a dialectical relationship between a particular discursive event and the situation(s), institution(s) and social structure(s) which frame it. A dialectical relationship is a two-way relationship: the discursive event is shaped by situations, institutions and social structures, but it also shapes them. To put the same point in a different way, discourse is socially constitutive as well as socially shaped.... Both the ideological loading of particular ways of using language and the relations of power which underlie them are often unclear to people. CDA [critical discourse analysis] aims to make more visible these opaque aspects of discourse.[25]

This perspective implies that the discursive struggle over participation is not a mere semantic struggle, but rather a struggle that is lived and practiced. In other words, participatory practices are—at least partially—structured and enabled by how we think of participation. The act of defining participation allows us to think, to name, and to communicate the participatory process (as minimalist or as maximalist) and is simultaneously constituted itself by our specific (minimalist or maximalist participatory) practices. As a consequence, the definition of participation is not a mere outcome of this political-ideological struggle, but an integrated and constitutive part of this struggle. In other words, the signification of participation is part of a "politics of definition,"[26] since its specific articulation shifts depending on the ideological framework that makes use of it. This also implies that debates on participation are not mere academic debates, but instead are part of a political-ideological struggle for how our political realities are to be defined and organized.

Finally, it should be emphasized that this struggle over participation is not exclusively embedded in the field of institutionalized politics. It touches all fields of the social, including the media field. Here we can return to Mouffe's concept of the political, which she sees as the "dimension of antagonism that is inherent to human society," in order to argue that the political touches upon our entire world, and cannot be confined to institutionalized politics.[27] The distinction Mouffe makes between the political and the social is also helpful because it locates the difference in the sedimented nature of practices. To use her words, this allows us to think of the struggle over participation as political, since it can potentially unsettle sedimented social practices:

> The political is linked to the acts of hegemonic institution. It is in this sense that one has to differentiate the social from the political. The social is the realm of sedimented practices, that is, practices that conceal the originary acts of their contingent political institution and which are taken for granted, as if they were self-grounded. Sedimented social practices are a constitutive part of any possible society; not all social bonds are put into question at the same time.[28]

A Second Coping Strategy: Taking Power More into Account

The second strategy to deal with the significatory diversity of participation distances itself from the question of a differentiation between authentic and pseudoparticipation. In contrast, this strategy focuses on the significatory process that lies *beneath* the articulation of participation and defines it as (part of) a political-ideological struggle. From this perspective, the definition of participation is one of the many societal fields where a political struggle is waged between different groups regarding the levels and intensities of their participation. Here the question is about the power positions of nonprivileged groups in democratic societies and attempts to strengthen (and weaken) their power positions, especially in regard to their power relations with privileged groups. Some models take a minimalist approach to participation, thus protecting the power base of privileged groups, while other models advocate maximalist participation as a means of equalizing the power relations between the different (privileged and nonprivileged) groups.

This struggle has manifested in democratic theory (and practice). More particularly, this struggle revolves around the always-present balance between representation and participation, or between the delegation of power and the exercise of power. This balance, for instance, provides structuring support for Held's typology in his *Models of Democracy*.[29] As Held describes it, "Within the history of the clash of positions lies the struggle to determine whether democracy will mean some kind of popular power (a form of life in which citizens are engaged in *self*-government and *self*-regulation) or an aid to decision making (a means to legitimate the decisions of those voted into power)."[30] Different models of democratic theory (and practice) strike different balances between the concepts of representation and participation. When the political is defined—following Schumpeter, for instance[31]—as the privilege of specific competing elites, which thus reduces the political role of the citizenry to participation in the election process, the balance shifts toward representation and the delegation of power. In this minimalist participatory-democratic model, societal decision making remains centralized and participation remains limited (in space and time). Moreover, in this model participation exclusively serves the field of institutionalized politics, as the political is seen to be limited to this field. In contrast, in other democratic models (for example, participatory or radical democracy),[32] participation plays a more substantial and continuous role and does not remain restricted to the "mere" election of representatives. These democratic models, which feature more decentralized societal decision making and a stronger role for participation (in relation to representation), are considered here to be maximalist forms of democratic participation. In these maximalist models, the political is seen as a dimension of the social, allowing for a broad application of participation in many different social fields, at both micro and macro levels, and maintaining a respect for societal diversity.

But the power struggle over participatory intensities is not limited to institutionalized politics within a democratic context. Often, these political practices diverge from "traditional" politics by aiming at cultural change; and sometimes several objectives and "targets" are developed in conjunction. For instance, the discourses of the feminist movement aimed at the rearticulation of gender relations and the empowerment of women within a diversity of societal spheres, and combined identity politics (see, for example, Harris's text)[33] with (successful) attempts to affect legal frameworks. Not only did this imply a broadening of the set of actors involved in political activities, it also incorporated an expansion of the spheres that were considered political. One example here is the feminist slogan "the personal is political," which claimed the political nature of social spheres such as the body and the family.[34] Millett, for instance, coined the term "sexual politics," thus extending the notion of the political into the sphere of the private and pleading for maximalist (female) participation in a range of societal spheres.[35] In her chapter on the "Theory of Sexual Politics," she introduces her sociological approach with the simple sentence, "Patriarchy's chief institution is the family."[36] A few pages on she notes that: "The chief contribution of the family in patriarchy is the socialization of the young (largely through the example and admonition of their parents) into patriarchal ideology's prescribed attitudes toward the categories of role, temperament, and status.... While we may niggle over the balance of authority between the personalities of various households, one must remember that the entire culture supports masculine authority in all areas of life and—outside of the home—permits the female none at all."[37] In these feminist projects we see (a plea for) the political (to) move further into the social, combined with the establishment of equal power relations among privileged and nonprivileged groups as its objective. We can apply a similar logic to other fields of the social, keeping Mouffe's argument in mind that the political touches upon our entire world and cannot be confined to institutionalized politics.[38]

Power, Participation, and Media: Another Example

These logics of struggle also apply, for instance, to the cultural/symbolic realm and the media sphere, and their political dimension should be acknowledged. In other words, the representational is political and an arena of the participatory power struggle. In this particular case, the struggle is waged between minimalist and maximalist models of media participation. In (very) minimalist forms, privileged groups, such as media professionals, retain strong control over process and outcome, often restricting participation mainly to access and interaction, to the degree that one wonders whether the concept of participation is still appropriate. Participation remains articulated as a contribution to the public sphere—but one which mainly serves the needs and interests of the mainstream media system itself, thus instrumentalizing and incorporating the activities of participating nonprofessionals. This media-centered logic leads to a homogenization of the audience and a disconnection of their participatory activities both from other societal fields and from the broad definition of the political, resulting in an articulation which posits media participation as nonpolitical. By contrast, in the maximalist forms, (professional) control and (popular) participation become more balanced, and attempts are made to maximize participation and to equalize the power relations between the different actors involved. This is often combined with an acknowledgment both of audience diversity and heterogeneity and of the political nature of media participation.

To zoom in more on mainstream media organizations, such as publishers and broadcasters, we see them acting as privileged and powerful signifying machineries that produce dominant representations and engage in the politics of representation (see, for example, Hall).[39] At the same time, they are revealed as organizational environments with specific politics, economies, and cultures where, for instance, the politics of the expert or the professional create power relations that impact upon the organization itself, but also on the "outside" world—acting as passage points through which this "outside" world is allowed in. Subject positions such as "journalist" and "media professional," but also "audience member," circulate widely in society, and carry specific—sometimes dominant—meanings that affect the position and power relations of the involved actors. The discursive affordances of these signifiers normalize, for instance, specific types of behavior, while disallowing other kinds of behavior. At the same time, these subject positions provide building blocks for people's subjectivities. Through the logics of identification, subject positions provide their opportunities for the exercise of agency. Here we should keep in mind that subject positions are not necessarily stable—and they can be contested, resisted, and rearticulated. The journalistic identity—and its articulation with professionalism—is worth mentioning here in particular for its combination of notions of public service, ethics, management of resources, autonomy, membership within a professional elite, the need for immediacy, and objectivity (see Deuze's and Carpentier's articles).[40] But the journalistic identity is only one among many subject positions that circulate within media organizations, and that contribute to the power struggle between minimalist and maximalist forms of media participation.

The specific positioning of (mainstream) media organizations within society strengthens their role as signifying machines. Obviously, media products have achieved a pervasive and spectacular presence in everyday life, to the degree that they have become difficult to (desire to) escape from. These media products are carriers of a multitude of discourses—contradictory ones, in many cases—but they do not always evade the workings of hegemony. In particular, their discourses about the media sphere itself often contain legitimizations for a media organization's hegemonic practices and cultures (see Couldry's book).[41] Among other functions, media products act as carriers of normalizing discourses about media organizations' claims to directly access reality, to centrality, and to an elite positioning in society. But, more importantly in the context of this chapter, they also include normalizations of mainstream media production cultures, where media professionals still hold strong—sometimes postpolitical—positions of power that enable them to internally manage the resources deemed necessary for providing publicness and visibility to, and framings for, other societal actors. Mainstream media and their media professionals do allow for specific forms of (what they label) participation—for instance, in formats such as audience discussion programs and reality television—but these formats severely restrict participatory intensities, rendering them as further examples of minimalist forms of participation (if they can be labeled participatory at all).[42]

In this sense, (mainstream) media organizations are machines that interrupt, channel, fixate, and produce flows. Their position also brings contestation, struggle, resistance, and instability, because the ways that they interrupt, channel, fixate, and produce flows are not always accepted. However dominant the mainstream media's organizational logics are, structural contestations of (some of) its basic premises remain, and these contestations (at least potentially, and often in reality) contain more maximalist definitions of participation. Through their discursive and material practices these contestations enter into a struggle with the more minimalist models of participation that mainstream media apply (and propagate).

One significant contestation is grounded in the sphere of alternative and community media organizations, which have introduced both a different model of media organization and a discursive attack on the privileged position of the media professional. Operating as civil society organizations, they are locations where internal participatory-democratic cultures and horizontal decision-making structures are frequently realized. In contrast to commercial and public broadcasters, these media organizations allow communities to participate in self-representational processes. These characteristics are nicely captured by Tabing's definition of the community radio station as "one that is operated in the community, for the community, about the community and by the community."[43] This participatory alternative/community media model emerged as a critical response to the internal logics of mainstream media organizations and their construction as large-scale, vertically structured, arbolic, sometimes bureaucratic organizations, staffed by professionals and geared toward large, homogeneous audience segments. Moreover, through this model mainstream media are critiqued as being carriers of dominant discourses and representations. The participatory alternative/community media model has not remained confined to its original habitats of print, radio, and (sometimes) television; it has also managed to find another home in the world of the online, triggering in the process a substantial increase of interest and hope for continued (media) democratization processes, alongside new incorporation strategies by mainstream media organizations. The affordances of online media, and some of its user cultures, have facilitated a nominal increase in discursive and materialist maximalist practices and projects; but at the same time, the online remains a prime location for even more mainstream minimalist-participatory and nonparticipatory media practices. Here we should bear in mind the long history (within both media and other social environments) of strategic use of the rhetorics of participation to protect a status quo in power imbalances.

Conclusion

The main argument of this chapter is that participation is not a fixed notion, but is nonetheless deeply embedded within our political realities, and functions thus as the object of long-lasting and intense ideological struggles. The search for harmonious theoretical frameworks able to capture contemporary realities might have been an important fantasy of *Homo academicus,* but it also may not have done any favors to measured attempts at analysis. This does not mean that conceptual contingency needs to be celebrated and radicalized; after all, "a discourse incapable of generating any fixity of meaning is the discourse of the psychotic."[44] It means, rather, that careful maneuvering is necessary to any project that would reconcile participation's conceptual contingency with the fixity needed to protect the concept from the signification of anything and everything.

By analyzing participation we come to see it as a contested notion, intimately connected with the power struggles in society. The disputation between minimalist and maximalist models of participation itself constitutes a power struggle over power—or, in other words, the debates surrounding minimalist and maximalist forms of participation are grounded in the political struggles regarding the power positions of privileged and non-privileged groups in society. The complexity, multilayeredness, and instability of these power struggles force us to move away from the ladder-based approaches—while still acknowledging their importance—and toward the analytics of participation, where the struggle between different actors over participatory intensities itself becomes an object of analysis. In the analytics of participation, minimalism/maximalism become a dimension, mappable onto different fields, spaces, and times—and sometimes transgressing them. In each of these (sub)fields, spaces, and times, a different balance between minimalism and maximalism can (potentially) be found. In this sense, minimalism/maximalism is more accurately multidimensional, and cannot be easily framed between the steps of one ladder.

Finally, it is important to emphasize that in the analytics of participation, minimalist and maximalist models of participation are understood to coexist in society. Although Mouffe's democratic revolution has intensified over time, it might be a bit early to celebrate the establishment of a participatory culture.[45] A subtle addition to Jenkins's terminology (he nowadays talks about "a *more* participatory culture,"[46] emphasis added) might bring us closer to an overarching analysis of contemporary power relations and the participatory condition. To quote Jenkins's argument at length: "Participatory culture, in any absolute sense, may be a utopian goal, meaningful in the ways that it motivates our struggles to achieve it and provides yardsticks to measure what we've achieved. More and more, I am talking about 'a more participatory culture' precisely to acknowledge this key principle—that participatory culture is something we have struggled toward over the past 100 plus years; we've gained ground and lost ground in and around each new technology."[47] Still, if we are to label the present-day (Western) participatory condition, within the context of the two-hundred-year-old democratic revolution, and at the risk of being speculative, I would like to suggest that it is more accurate to label it the era of minimalist participation.

Notes

1. Carole Pateman, *Participation and Democratic Theory* (Cambridge: Cambridge University Press, 1970), 1.

2. This chapter builds on the reflection on participation developed in Nico Carpentier, *Media and Participation: A Site of Ideological-Democratic Struggle* (Bristol: Intellect, 2011).

3. Sidney Verba, *Small Groups and Political Behavior* (Princeton, N.J.: Princeton University Press, 1961), 220–21.

4. George Strauss, "An Overview," in *Organizational Participation: Myth and Reality,* ed. Frank Heller, Eugen Pusic, George Strauss, and Bernhard Wilpert (New York: Oxford University Press, 1998), 18.

5. Sherry R. Arnstein, "A Ladder of Citizen Participation," *Journal of the American Institute of Planners* 35, no. 4 (1969): 216–24.

6. Ibid., 216.

7. Ibid.

8. Pateman, *Participation and Democratic Theory*, 70.

9. Ibid., 71.

10. Stephen Ward, *Ethics and the Media: An Introduction* (Cambridge: Cambridge University Press, 2011).

11. Ibid., 227.

12. Henry Jenkins and Nico Carpentier, "Theorizing Participatory Intensities: A Conversation about Participation and Politics," *Convergence: The International Journal of Research into New Media Technologies* 19, no. 3 (2013): 275.

13. Janet Wasko and Vincent Mosco, eds., *Democratic Communications in the Information Age* (Toronto: Garamond, 1992), 7.

14. Peter Dahlgren, *Media and Political Engagement* (New York: Cambridge University Press, 2009); Nico Carpentier, Kim Schrøder, and Lawrie Hallett, "Audience/Society Transformations," in *Audience Transformations: Shifting Audience Positions in Late Modernity,* ed. Nico Carpentier, Kim Schrøder, and Lawrie Hallett (London: Routledge, 2014), 10; Carpentier, *Media and Participation.*

15. Alberto Melucci, *Nomads of the Present: Social Movements and Individual Needs in Contemporary Society* (Philadelphia: Temple University Press, 1989), 174.

16. Anthony Giddens, *Runaway World: How Globalisation Is Reshaping Our Lives* (London: Profile, 2002).

17. Hubert L. Dreyfus and Paul Rabinow, *Michel Foucault: Beyond Structuralism and Hermeneutics* (Chicago: University of Chicago Press, 1983), 186.

18. Gavin Kendall and Gary Wickham, *Using Foucault's Methods* (London: Sage, 1999), 50.

19. Michel Foucault, *History of Sexuality, Part 1: An Introduction* (New York: Pantheon, 1978), 94.

20. Ernesto Laclau and Chantal Mouffe, *Hegemony and Socialist Strategy: Towards a Radical Democratic Politics* (London: Verso, 1985).

21. Chantal Mouffe, *On the Political* (London: Routledge, 2005), 9.

22. Chantal Mouffe, *The Return of the Political* (London: Verso, 1997), 3.

23. Mouffe, *On the Political,* 18.

24. Laclau and Mouffe, *Hegemony and Socialist Strategy.*

25. Norman Fairclough and Ruth Wodak, "Critical Discourse Analysis," in *Discourse as Social Interaction,* ed. Teun A. van Dijk (London: Sage, 1997), 259.

26. Katherine Fierlbeck, *Globalizing Democracy: Power, Legitimacy and the Interpretation of Democratic Ideas* (Manchester: Manchester University Press, 1998), 177.

27. Mouffe, *The Return of the Political,* 3.

28. Mouffe, *On the Political,* 17.

29. David Held, *Models of Democracy*, 2nd ed. (Cambridge: Polity Press, 1996).

30. Ibid., 3. Emphasis in original.

31. Joseph Schumpeter, *Capitalism, Socialism and Democracy* (London: Allen and Unwin, 1976).

32. Pateman, *Participation and Democratic Theory*; C. B. Macpherson, *Democratic Theory: Essays in Retrieval* (Oxford: Clarendon Press, 1973).

33. Duchess Harris, "From the Kennedy Commission to the Combahee Collective: Black Feminist Organizing, 1960–1980," in *Sisters in the Struggle: African American Women in the Civil Rights-Black Power Movement*, ed. Bettye Collier-Thomas and V. P. Franklin (New York: New York University Press, 2001), 280–305.

34. Carol Hanisch, "The Personal Is Political," in *Notes from the Second Year: Women's Liberation, Major Writings of the Radical Feminists*, ed. Shulamith Firestone and Anne Koedt (New York: Radical Feminism, 1970), 76–78.

35. Kate Millett, *Sexual Politics* (Garden City, N.Y.: Doubleday & Company, 1970).

36. Ibid., 33.

37. Ibid.

38. Mouffe, *The Return of the Political*.

39. Stuart Hall, "The Spectacle of the Other," in *Representation. Cultural Representations and Signifying Practices*, ed. Stuart Hall (London: Sage, 1997), 257.

40. Mark Deuze, "What Is Journalism? Professional Identity and Ideology of Journalists Reconsidered," *Journalism* 6, no. 4 (2005): 442–64; Nico Carpentier, "Identity, Contingency and Rigidity: The (Counter-)Hegemonic Constructions of the Identity of the Media Professional," *Journalism* 6, no. 2 (2005): 199–219.

41. Nick Couldry, *Media Rituals: A Critical Approach* (London: Routledge, 2003).

42. The participatory intensities in reality television, especially, are problematically low (see Mark Andrejevic, *Reality TV: The Work of Being Watched* [Oxford: Rowman & Littlefield, 2004]). But even in audience discussion programs the levels of participation tend toward the more minimalist versions (see Nico Carpentier, "Managing Audience Participation," *European Journal of Communication* 16, no. 2 [2001]: 209–32).

43. Louie Tabing, *How to Do Community Radio: A Primer for Community Radio Operators* (New Delhi: UNESCO, 2002), 9.

44. Laclau and Mouffe, *Hegemony and Socialist Strategy*, 112.

45. Chantal Mouffe, *The Democratic Paradox* (London: Verso, 2000).

46. Jenkins and Carpentier, "Theorizing Participatory Intensities," 266.

47. Ibid.

CHAPTER 2

Participation as Ideology in Occupy Wall Street

Cayley Sorochan

NO MATTER WHERE IT APPEARS, participation consistently carries with it a positive connotation. In many cases the positive connotation attached to participation implies that it is a good in itself, regardless of the nature of the practice we are encouraged to take part in. Nevertheless, participation is a contradictory concept. On the one hand, it promises collective emancipation and individual self-empowerment; on the other, it incorporates energy, creativity, and labor into the reproduction of existing institutions and power structures. This ability to circulate widely and compel affirmative responses indicates that participation has attained the status of ideology. I approach participation as an ideological concept by mapping out the implicit values invoked by the term. The "participatory complex," as I define it, may include: a valuation of activity over passivity; the privileging of procedure or structure over ends; a desire for immediacy and antirepresentational attitudes; the privileging of face-to-face encounters or bodily copresence; an orientation toward inclusiveness and pluralism; a will to consensus; and discourses of empowerment through personalization. Whenever participation is proffered or demanded, some combination of these values is invoked to galvanize desire.

The typical response to the contradiction between the democratic promise of participation and its use by nondemocratic institutions is to maintain participation as an ideal by distinguishing *genuine* participation from the mere appearance of involvement. Yet even genuine practices of participation exhibit internal limitations and contradictions. It is my contention in this essay that the limitations of participation cannot be addressed solely by adjusting decision-making structures or procedures to ensure that they are "more participatory"; rather, the very ideal of participation itself must be questioned and tempered in relation to other more substantial political goals and principles. What concerns me are the moments

when participation is accepted as an ideology, and the negative consequences this may pose for radical organizing at the level of political subjectivity. I am less concerned with the internal procedures of participatory
decision making, which vary widely and differ in effectiveness from context to context. It is therefore important to differentiate between democratic decision-making structures, with which participation is often conflated, and participation as an ideology that consists of values that exceed
these practices.

Since 2011, there has been a wave of global activism critical of growing
inequality, neoliberal governance, and the undermining of democratic
institutions.[1] Direct democratic practices have been a central part of
movements such as the Indignados or M15 movement in Spain, the anti-
austerity movement in Greece, and Occupy Wall Street (OWS) in the
United States. For OWS in particular, participation not only inspired a
method of decision making but was also held up as a central ideal and
goal. However, in a context where participation already enjoys hegemonic
status as a value in itself, to what degree and in what way can participation
challenge existing power structures? I will argue that participatory democracy is an ideology that directly challenges the inequity of representative
political systems only when it entails a radical economic transformation
based on an egalitarian rejection of capitalist property rights. Moreover, the
educative, cooperative, and integrative values that participatory democracy
foregrounds are not adequate to the confrontational class politics such
transformation implies if they are not accompanied by strong principles of
solidarity, collectivity, and clarity over shared interests.

In the following I examine the meaning of participatory democracy in
OWS and attend to how certain values attached to participation—including
valuations of structure or process over ends, an absolutist interpretation of
inclusiveness, and antirepresentational attitudes—formed an ideological
complex that stood in the way of functional organization. Within OWS
there were many who were critical of capitalism, and the movement drew
attention to the unfairness of the existing system and the greed of the one
percent. At the same time, an abstract and apolitical version of participation, as seen in the official statements and declarations of the movement as
well as in some of its practices, came to trump this critique. Participation
was upheld as an ultimate value regardless of whether it contributed to
building democratic power oriented toward an egalitarian transformation
of capitalism or undermining it. Insofar as Occupy registered a shift in

political horizons, it is due to its success in asserting a politics of class division and building collective power behind the campaigns of its working groups. However, at the level of its highest decision-making body, occupiers refused to draw boundaries based on progressive political principles or ideological coherence and instead insisted that an organizational structure based on inclusion and pluralism was adequate to the radical transformation that it hoped for.

The Radical Implications of Participatory Democracy

Participatory democracy is the very first of the principles of unity and solidarity declared by the OWS New York general assembly on September 23, 2011.[2] But what is meant by participatory democracy? Movement theorists and practitioners typically define participatory democracy in opposition to representative democracy and contend that participation is the necessary characteristic of a genuine democracy. For example, Carol Pateman describes a split in classical democratic theory between those who consider participation to have a narrow protective function and those who understand participation to be fundamental to democratic culture.[3] For James Mill and Jeremy Bentham, the people's participation in politics through voting and discussion is an important check on government power because it is thought to ensure protection of the private interests of each citizen and therefore further the universal interest.[4] Here, democracy is understood merely as a method of choosing leaders rather than a system that develops citizens' capacities for self-government. In contrast, Pateman describes how theorists such as Jean-Jacques Rousseau, John Stuart Mill, and G. D. H. Cole place a much stronger emphasis on the role of participation in democracy.[5] For these thinkers democracy does not refer simply to a system of government but to a broad democratic culture wherein participation is extensive and habitual. The ultimate aim of participatory democracy is not only the protection of individual freedoms, but also the development of uniquely human qualities, capacities, and potential.

Pateman ascribes three primary functions to participation in theories of participatory democracy. The first and most fundamental role of participation is education.[6] In order for citizens to develop the necessary democratic attitudes, skills, and psychological qualities that would allow for and sustain a meaningful democracy, they must have access to democratic spaces where these traits can be learned and exercised. For instance, she points to Rousseau's argument in *The Social Contract* that democratic skills and attitudes need not preexist the space of participation because the participatory process itself is designed to develop individuals' capacities for responsible social and political action.[7] Rousseau thought that in a deliberative space where every participant enjoys equal influence, no single individual would be able to impose inequitable demands on the rest. Individuals would therefore be compelled to take into consideration the broader public interest if they were to gain the cooperation of others.[8] Participation is therefore thought to be broadly educative, in that it develops a sense of collective justice as well as the skills required to articulate its meaning. The second function of participation described by Pateman is that it enables collective decisions to be more easily accepted by the individual, enhancing cooperation. This is because participation offers the individual opportunities to engage in and influence the outcome of decisions, and because it fosters interdependence among participants.[9] Participation also has an integrative function; taking part in collective decisions helps individuals to feel like they are a part of the community and deepens their sense of attachment to society.[10]

In order for participation to effectively play the roles ascribed to it, access to participatory decision-making processes must be far more widespread than it is under existing electoral regimes. Participatory democrats contend that participation in the political sphere would be supported and enhanced by participation in other social institutions, such as education, culture, the family, and the workplace.[11] This last institution is particularly important to Pateman as it is both the place where adults spend most of their time and a primary site where local relations of authority condition workers into roles of subservience.[12] Furthermore, all of the theorists of participatory democracy recognize that for it to function as desired, a substantive degree of economic equality is a necessary condition. Formal participation in decision-making structures is not enough to guarantee an equal share in power over decisions. Every participant must enjoy a high measure of economic security and independence if they are to maintain their autonomy in the decision-making space and if they are to enjoy a roughly equal ability to develop their human capacities. Rousseau, for instance, based his theory of participation on an economic system of small peasant proprietors, one where "no citizen shall be rich enough to buy another and none so poor as to be forced to sell himself."[13]

For most classical liberal democratic theorists, property rights were held to be a fundamental guarantee of individual independence and freedom. According to C. B. Macpherson, what separates a thinker like Rousseau from others like James Mill, Bentham, and J. S. Mill is that these latter thinkers all attempt to reconcile the egalitarian ideals of democracy with a continued acceptance of capitalist market relations. Macpherson argues that in a market society where unlimited individual appropriation of property is allowed, resulting class divisions undermine the egalitarian principles that serve as a moral justification for this system.[14] This contradiction between democratic ideals and class division is particularly acute in relation to what Macpherson calls theories of "developmental democracy."[15] Political thinkers such as J. S. Mill, A. D. Lindsay, and John Dewey primarily understood democracy to be a means of individual self-development. They believed that the existing democratic system of their time was, alongside education, communication, and social knowledge, the means to human progress.[16] They were dismissive of the significance of class division, arguing that class was being replaced by pluralistic social groups and that the redistribution of wealth would occur gradually through voluntary cooperation between employers and workers.[17] Macpherson warns that recent attempts to revive participatory democracy, such as those made by the New Left of the 1960s, must be careful to not repeat the mistakes of these thinkers.[18]

Participatory democracy is a radical politics not only because it entails widespread changes in social relations within the political sphere, but also because in order to function as desired it entails fundamental economic transformation. For this reason, participation may be an attractive if somewhat deceptive way to sneak in an anticapitalist politics under the broadly uncontroversial mantle of "democracy." In a cultural context like the United States, where democracy is ideologically equated with free markets, and socialism is often equated with Stalinism or totalitarianism, participation may seem to be the most effective way to get people to question current economic arrangements.[19] However, the prospect that economic equality should be considered a condition of meaningful participation raises the question as to how the transformation away from a vastly unequal capitalist economy can be achieved and whether the idea and practice of participation on its own is enough to effect such a transformation. The values of education, compromise, integration, deliberation, individual empowerment, and self-development that are central to participatory

democracy in a postcapitalist context do not necessarily point a way forward for a struggle that necessitates a confrontational politics, a commitment to collective interests, and a willingness to engage in divisive class
conflict in the present. To rely on the former set of values alone would be
to depoliticize this struggle and bring it too close to the progressive but
unrealistic presumptions of the developmental democrats. On the other
hand, participatory relations could also be put in the service of confrontational political struggle if the educative, cooperative, and integrative
functions it performs are used to mobilize a broad base of the population
behind a common political goal that is understood to exist within a field of
deeply opposed interests.

OWS, a movement that centered itself on participatory democracy and
was motivated both by a critique of money in politics and by growing
inequality, would seem to be the perfect example of a radical and participatory politics aimed at challenging capitalism at a systemic level. However, as I will argue below, the values that participation was imbued with in
OWS came to undermine its attempts to build a radical movement for
social change.

Occupy Wall Street: Genuine Democracy or Class Division?

Occupy Wall Street was the most successful leftist social movement in the
United States since the Global Justice Movement of the late 1990s. What
began as a small encampment of a few dozen people ballooned within a
week into a camp of six hundred. Within a month it became a nationwide
and even global movement as similar camps were set up in over five hundred cities. Something about Occupy clearly struck a chord. Although
the camps themselves were violently evicted in November 2011 in what
appeared to be a nationally coordinated effort by police to repress the
movement, and although the movement never did determine a unifying
demand or make concrete gains, it has been argued that OWS had a lasting ideological effect on Americans' political horizons. David Graeber, an
anarchist scholar and one of the original organizers of OWS, argues as
much in his book *The Democracy Project,* where he emphasizes the profound opening of the radical imagination that the movement produced.[20]
Graeber especially privileges the experience of deliberative democracy
that participants encountered in the Occupy general assemblies (GAs) as
key to this transformation of the imagination.

The camp that was set up in New York City's Zuccotti Park on September 17, 2011, began with a GA. Its initial purpose, based on the question posed by *Adbusters* magazine, which launched the "Occupy Wall Street" meme in April, was to determine the movement's one demand. However, as the assemblies progressed, the OWS camp avoided making a single demand, and this avoidance strategy became one of its central characteristics. Even without a single demand, and in spite of the media's disingenuous confusion about the movement's purpose, the main concern of OWS was clear: the power of corporations to undermine genuine democracy in America.[21] The economic crisis of 2008 and the government bailout of the nation's banks emphasized the stark inequalities that had long become the norm.[22] The individuals responsible for the crisis remained unpunished while the homes of the poor were foreclosed. The lack of justice was acute, but whether or not it would lead to a movement for economic justice based in redistributive politics or a critique of capitalism in its entirety remained to be seen.

The alternative to the systemic domination of politics by corporations that was proposed by the movement turned out to be neither a welfare state redistribution of wealth nor the socialization of property, but rather participatory democracy as modeled by the GA. At OWS the daily GA was the highest decision-making body of the movement. The work of maintaining the camps, organizing direct actions, producing research, doing media outreach, and resolving all kinds of internal problems was delegated to voluntary working groups (WGs), which were supposedly approved by and answerable to the GA. The GAs were formally open to anyone present and operated through a consensus process. The procedures of the GAs were widely publicized and included a system of hand signals. Further, the "mic check" technique, wherein GA participants would collectively echo the words of a speaker, was used to amplify single voices for a large crowd. The WGs of the New York occupation organized various marches in collaboration and solidarity with various community groups and labor organizations. They also spent plenty of time organizing the logistics of the camp and defending it against political, legal, and police repression. Beyond the direct actions and the pragmatic concerns produced by the tactic of occupation, the WGs engaged in discussions oriented toward confronting the issues of corporate power and democracy that had attracted people to the GAs in the first place. A substantial amount of documentation and proposals came out of these WG discussions, and they reflect a diverse set of concerns—from the housing crisis and student debt and unemployment, to campaign finance reform, alternative economics, and the NYPD's Stop and Frisk program.[23] While the WGs generated a number of actions and campaigns, some of which continue to this day, the diversity of issues they addressed made achieving consensus within the GA on many WG proposals extremely difficult.

Marina Sitrin, one of the initial organizers of OWS, claims that what was most important in this movement was opening up a space for conversations—for "real, direct, and participatory democracy."[24] Rather than make demands of those in power, the assembly movement sought to strip the existing system of legitimacy by creating the most open and participatory space possible. This space was intended to stand in stark contrast to the broken system of parliamentary democracy that was incapable of defending the public interest against the interests of the one percent. The avoidance of presenting a single demand to those in power underlined the movement's autonomy and made it clear that OWS aimed at a deeper, more systemic challenge. But to what degree or in what sense is the process of deliberation, conversation, and participation disruptive to the existing power structures or to the dominant ideology of democracy in America?

In his analysis of Occupy, Graeber contrasts the democracy of public assemblies during the American Revolution, in South African and Indian village councils, on pirate ships during the colonial era, and in the general assemblies of OWS with the republican system of elite power that is called representative democracy.[25] He argues that democracy has always been a destabilizing concept, threatening to those in power due to its rejection of systems of hierarchy. Essentially, for Graeber, OWS reconfigured political horizons by offering an example of genuine democracy in contrast to the false democracy of the existing system. Yet Graeber pays relatively little attention to the way democratic power can effect changes in the face of antidemocratic forces, focusing instead on the functionality of participatory processes and the noncoercive ways that public assemblies can come to collective decisions. But then, how does such a democratic space defend and extend itself? The only option appears to be the power of persuasion that can result from setting an example. This would belie, however, the real collective power that supported the existence of the camps in the form of mass mobilization.[26] Clearly, democratic assemblies are a threat to inequitable systems because they have the potential to unleash a collective power that would actively contest elite rule. The internal functionality of a participatory process demonstrates the possibility of organizing political life differently, but the power to compel such reorganization lies in sustained direct action.

Jodi Dean offers an interpretation of OWS that locates its most compelling dimension in the naming of a fundamental division and an affirmation of collective power.[27] She insists that what was radical about OWS was not that it embodied genuine democracy, but rather its insistence on a politics of division that was encapsulated in the slogan: "We are the ninety-nine percent." This statement politicized an economic fact and in the process produced a collective identity based on a class relation rather than substantial identities based in race, nationality, or religion.[28] OWS should therefore be seen in contrast to the kinds of political gestures that presume society to be simultaneously a coherent unity and a pluralistic field of difference.[29] Dean argues that the collective subject that emerged in this political rupture cannot be reduced to the real multiplicity of individuals and opinions that were integrated through the democratic process. The collective is not simply an expression of an aggregate of these individuals but an idea that transforms its participants by changing how they relate to their setting.[30] Dean characterizes the relation that OWS adopted as one of antagonism toward processes of proletarianization. For Dean, it was the confrontational tactic of occupation and not the GAs per se that were effectively threatening to those in power.[31] The difference in emphasis between Graeber's focus on internal process and Dean's focus on class consciousness illustrates the tension that existed within Occupy's own self-understanding. I would argue that the significance of the camps was the space they opened up not for conversation but for organization, and this is what the ideological emphasis on participation occludes.

Participation as Ideology

It is important to insist on the difference between direct democratic practices and the set of values that shape the ideology of participation. If a problem of participation is apparent, the solution is sought out through continuous procedural amendments that seek to establish the perfect and unassailable organization that would live up to our conception of a true, genuine democracy. Yet it may be the case that the source of conflicts and inconsistencies lies within the very ideal against which we would seek to measure our practices. If participation is interpreted to mean total inclusion, a very high degree of individual autonomy, a high value placed on self-expression, and an avoidance of divisive political stances, it is possible to see how this set of values may in fact contradict or stand in the way of meaningful democratic practice. Attending to this division between democratic practice and an ideology of participation allows us to imagine an organization that functions democratically while it nevertheless foregrounds other core principles and substantive political aims beyond the space of deliberation. Second, it helps to clarify how the problems of participation may not be solved by procedural changes alone but also require that we confront the ideological blockage that lies at their source. In some cases a charge that a process is not participatory enough may disguise a more substantive political difference.

One example of an approach to participation that places a bit too much weight on procedures can be found in Graeber's emphasis on consensus as the source of democratic cultural transformation. In a fashion similar to proponents of participatory democracy, Graeber argues that achieving a genuine democracy entails more than a change in the structure of government; it requires the creation of a democratic culture. He likens the potential shift to a deep moral transformation similar to the kind of cultural change initiated by feminism in the 1970s.[32] For Graeber, consensus is transformative because it is based on listening and compromise rather than a contest between fixed and preexisting interests. However, L. A. Kauffman points out that it is precisely the assumption "that division results from differing views (which can be reconciled) rather than competing interest (which often cannot)" that leads consensus to be adopted with an uncritical reverence by many segments of the activist left.[33] Kauffman is concerned that consensus, as derived from the Quaker tradition, has become an article of faith within movements like Occupy. Rather than a matter of faith, we could just as easily understand the absolutist interpretation of participation to be a matter of ideology.

In chapter 4 of his book, Graeber addresses a number of perceived problems or criticisms of the consensus process.[34] It is interesting that Graeber regularly dismisses failures of process as resulting from a lack of experience or improper use of procedures. This is his rationalization for the problems of historical uses of the consensus process in Students for a Democratic Society, the antinuclear movement of the 1970s, and also the problems of consensus as used by OWS. While it is certainly true that few people have extensive experience with any kind of direct democratic decision-making process and that this kind of experience is crucial, the examples of failures of process in OWS that Graeber raises are more illustrative of an absolutist interpretation of participation than they are of the authoritarian or hierarchal attitudes that he associates with majoritarian methods of democracy. It is more pressing that we differentiate between democratic culture and participatory culture than that we focus on distinguishing democracy from authoritarianism. I will explore a few of the examples he raises in the remainder of this essay.

Structure/Process/Procedures over Ends

A typical trait of participation as ideology is a fetishization of organizational structures and processes as goals in themselves. When participation operates ideologically, the particular organizational structure or process, if properly adhered to, is thought to guarantee a beneficial or legitimate outcome, or the process or structure itself is understood to be more important than any eventual decision or consequence. A confusion of means with ends can follow from an obsession with process and result in the undermining of effective practice oriented toward bettering social conditions.

As mentioned above, in OWS the GA and its procedures were not regarded as a mere method of collective decision making; they were held up as vitally important to the meaning of the movement itself. Initially, the satisfaction of achieving consensus on anything among a very diverse group of participants and the necessity of maintaining the camp was enough to propel continued participation. However, eventually the fetishization of the GA and the consequent inability to cohere around a broad actionable idea of social change beyond the GA itself undermined people's commitments to the space. The firsthand anecdotes from the *Occupy* collection register a shift from feelings of exhilaration to feelings of frustration and futility. For instance, Keith Gessen describes the underwhelming experience of a Russian friend who, after participating in the New York GA, described somewhat cynically how he "managed to catch the entire discussion about whether they [the GA] should buy some shelves."[35] Regardless of whether they are participatory or not, for many people, meetings are tedious affairs. The purpose of engaging in them is not to achieve a world in which we spend all of our time in GAs, but to achieve a more equal and just society. While these ends cannot be neatly separated from the process, neither can they be completely collapsed. The meaning and efficacy of democratic spaces are dependent on their ability to build and channel collective power. Agreeing to achievable goals external to deliberation itself contributes to the momentum required to maintain democratic institutions.

In Graeber's guide to consensus, he repeatedly acknowledges the need for a democratic group to come together around common goals, principles, or interests.[36] He also claims that these principles of unity and common goals should be kept simple and general. Yet there is a difference between simplicity and abstraction. OWS's lack of ideological coherence and extreme openness contradicted this warning. This can be seen most clearly in the principles of the solidarity document, which was one of the few statements to be formally adopted by the GA during the occupation.[37] This document was intended to be a "living document" that would develop over time. However, the fact that the very basis of the existence of the group was to be clarified by the group through consensus points to a tautological problem. Those attempting to determine the basis of their collectivity did not have an agreed-upon basis according to which their attempts at consensus could be measured. It is therefore unsurprising that the core principles of solidarity intended to define the purpose and contours of OWS are self-reflexive, vague, and so general as to risk undermining the critical impetus that drew people to OWS in the first place.

A self-reflexive tendency is apparent in the way that most of the principles attend to the internal relations of the group, rather than the common experiences or interests that have brought them together. The first principle is procedural and declares an intention to engage in "direct and transparent participatory democracy." The second, "exercising personal and collective responsibility," is meaningless without further elaboration, as is the fifth: "redefining how labor is valued." The third and fourth principles contain references to privilege and oppression, but the emphasis in both cases is on individual interactions and personal empowerment, with the third recognizing an "individual's inherent privilege and the influence it has on all interactions." The fourth is "empowering one another against all forms of oppression." In these two cases, rather than refer to concrete historical systems of discrimination and inequality, such as a clear rejection of colonialism, sexism, or class inequality, the substantial political content is depoliticized through generalization. Furthermore, while the concept of privilege is intended to make us sensitive to the power differentials that exist within the ninety-nine percent, focusing too much on individual interactions could end up undermining our ability to act collectively to confront these inequalities. If social change is seen to rest entirely on individual moral transformation, meetings can degenerate into moralizing critiques when the preferred outcome would be solidaristic collective action against, for instance, racialized policing policies. The remaining two principles, "the sanctity of individual privacy" and "the belief that education is a right," are basic and uncontroversial liberal assumptions formally endorsed by the United Nations and most liberal democratic states.

Nowhere do these principles of solidarity or points of unity describe a common condition of injustice or exploitation, nor do they gesture toward a positive vision of collectivity. For instance, given the list of grievances that OWS released in its declaration, it is surprising that the GA did not see fit to find some basis of unity in defense of the commons or an expansion of the public sphere. The principles of unity passed by the New York GA are so open to interpretation that anyone could feel comfortable agreeing to them. They avoid any divisive political content that would clearly demarcate this movement from reactionary, populist, individualistic, and market-based ideologies. This is not true, of course, of the movement as a whole; its concept of the ninety-nine percent asserted a degree of class consciousness, and its various working groups exhibited a strong degree of political consciousness in relation to social and economic inequalities. The WGs in many cases created documents that contained rich supporting detail and numerous concrete proposals for action. However, the respective proposals of different working groups were often derived from extremely divergent political assumptions. Consequently, few proposals managed to achieve consensus at the GA during the occupation, aside from those that dealt with the pragmatics of daily existence.

Inclusivity or Openness

One of the most difficult obstacles confronted by the OWS GA in its attempts to come to substantial and actionable collective decisions was its policy of inclusiveness. Inclusivity is a basic principle of any direct democratic process because the very concept of direct democracy indicates an opening up of decision-making spaces to those with some kind of concrete stake in the issue at hand. Inclusivity becomes a dimension of participation as ideology when it is treated as a value independently of its basis in a specific collective problem, or when the purpose or common interests of the group are poorly defined.

When discussing the issue of implicit and explicit boundaries to participation, Graeber describes one particular conflict in a New York Spokescouncil meeting that to my mind is indicative of the absolutist approach to inclusion that characterized OWS from the beginning:

> I recently attended a Spokescouncil in New York where everyone had been engaged in a long debate over whether there should be a "community agreement" and a shared principle that if anyone violates that agreement, they should be asked to voluntarily leave. The proposal was meeting concerted opposition when, suddenly, someone noticed one of the delegates was holding a plaque saying "Aryan Identity Working Group." He was immediately surrounded by people—many of those who had just been loudly insisting such a rule was oppressive—who successfully forced him to leave.[38]

Graeber raises this example for the sake of insisting that even inclusive groups need to maintain certain boundaries. What interests me here is not the fact that a community agreement was ultimately passed, but that the very idea of enforcing boundaries on the movement was seen as "oppressive" at all. Graeber also notes that during the occupation even when people would declare that their purpose was to disrupt a meeting they were nonetheless allowed to take part, but he does not comment on the ideological premises of this absolutist attitude toward inclusion.[39] These are clear instances where that which stands in the way of democratic practice is not a readiness to issue commands or inexperience with democratic process, but an unwillingness on an ideological level to politicize the boundaries of the movement. This rejection of politicization is enforced by the belief in participation as an ultimate principle above all others.

Antirepresentational Attitudes

The aim of participatory democracy is to disperse decision-making power as widely as possible so that everyone has an equal influence over matters of public concern. Representative power, understood as a temporary decision-making authority vested in an individual or group, is therefore limited as much as possible to avoid unequal concentrations of power and the possibility of corruption. For coordination at higher levels, or for the execution of the collective will, delegates are typically preferred over representatives. Delegation is a more limited form of representation in which a person is authorized to present the views of a group to others or to act as they are mandated, but not to make independent decisions that contradict these views or mandates. A delegate can be recalled at any time, and any agreements made by a delegate with other parties are not considered legitimate until the local group validates them.

Both Pateman and Macpherson acknowledge the need to delegate power and even presume the necessity of a minimal amount of representative power in the context of large-scale industrialized nations. Pateman relies on the theories of guild socialism developed by G. D. H. Cole in order to clarify the acceptable role of representative power in a participatory democracy. According to Cole, democracy is only real when it is conceived in terms of the principle of function and purpose.[40] The problem with existing forms of representative power under the parliamentary system is that it is assumed that an individual voter can be adequately represented on every issue. After the representatives are elected, the electors no longer have control over them and cannot recall them. In contrast, under a system of guild socialism that is built on functional associations that operate through participatory democracy, representatives would have a far more limited and precise mandate and would be subject to continuous advisement, criticism, and the possibility of immediate recall by the local association that elected them.[41] In this context of widespread participation, where executive power is based on specific functions and subject to popular oversight, Cole did not think that limited amounts of representative power would undermine the equality of participation.

It is evident that many participants in OWS exhibited an extreme aversion to representation that goes far beyond the critique developed by participatory democrats. This reflects a longer trend that results from the anti–"big government" campaigns of the right since the 1980s, as well as the rejection of hierarchy within the New Left social movements of the 1960s. A total rejection of representation leads to a paradoxical situation: The initiative and autonomy of each participant can be severely limited if, in an effort to avoid "representing" anyone else, they believe they do not have the authority to act on behalf of or in the name of the collectivity without constant and explicit approval. This was the case in an example given by Graeber which occurred in a GA prior to the establishment of the camp. A conflict emerged when the Outreach WG brought to the GA for approval a two-line description of the nature and purposes of the Occupy movement to be used in flyers.[42] The statement was met with continuous objections no matter how minimal and uncontroversial the wording chosen by Outreach. Graeber argues that the problem was the presumption on the part of the WG that they needed approval from the GA for such minute details; they had already been empowered to do outreach. This would therefore be an instance where an antirepresentational principle was inflexibly applied to the most basic of tasks, thereby undermining the autonomy of what we could consider de facto executants or representatives of the collective will. However, I would argue that Graeber's solution in this case to simply assume the power to act without gaining further approval is also inadequate if it operates outside of a formal means of accountability on the part of WGs to the GA. The fact that members of the WGs were unelected volunteers and could not be recalled or expelled by the GA does open up the potential for abuse (or in this case, the inability to act and feelings of frustration and demobilization) to an even greater extent than the limited representative power described by Cole. Graeber's argument that formal power is somehow more pernicious than informal power is not convincing as he conflates all "representative" power with the traditional form criticized by participatory democrats and does not distinguish this from more limited, recallable, and local forms. There is also no good reason offered as to why the existing powers of WGs as de facto executives could not be more clearly defined and limited, or why a process of recall could not be instituted.

Conversely, rather than an overly subservient relationship to collective oversight, the rejection of representation can also lead to an extremely individualistic or atomistic refusal of any kind of collective authority. We might understand the overreadiness to "block" proposals that one only mildly disagrees with as an example of this. Graeber insists that blocks should only be used in cases where a proposal contradicts a group's fundamental principles of unity or shared purpose.[43] Here I would again assert that the extremely vague principles of OWS made it difficult for participants to agree on which blocks were appropriate. A more specific example could be found in the controversy that emerged when the drum circle was asked to limit its playing to certain hours.[44] Apparently impervious to the way that the drums were undermining community support, the ability of the movement to communicate, and the comfort of campers in already difficult circumstances, the drummers defended their resistance to these requests by arguing that they had a right to freedom of expression. One drummer was quoted on NPR saying, "They've [those in the camp who opposed their drumming] turned into the government that we've been trying to protest!"[45] Here is an instance where those participating in the camps did not necessarily participate in the GAs or recognize the authority of that body, instead understanding the camps as a space of individual or subcultural expression that recognized no higher level of representation. With no will to compel the drummers to stop, the "solution" to this problem was essentially to ignore it.

Occupy or Reject Power?

OWS signaled a powerful rejection of corporate power, class inequality, discrimination, and the representative system of government. When it came to articulating a positive vision of social transformation, the movement relied on an ideology of participation. The result of this ideology of participation was a lack of shared political principles, an overemphasis on organizational structure as the key to equality, and antirepresentational attitudes. All of these traits speak to a deep rejection of power. While the desire was to undermine illegitimate hierarchies, it also led to an inability to recognize and catalyze legitimate democratic power, insofar as the GA was tentative in defining and defending the boundaries of the decision-making body with reference to common interests and political goals. The GA balked at politicizing its boundaries because doing so would entail making the kinds of strong political statements that would lead many people to exclude themselves from the movement. Participation therefore acted as the fulcrum on which competing definitions of democracy rested. On the one hand democracy was interpreted as an inclusive and pluralistic process of deliberation; on the other it was an antagonistic practice oriented toward channeling collective power in the direction of class struggle.

The most lasting element of OWS was not the space of the GA but the campaigns started by some of the WGs. Their more limited functions made them more ready to assume the power to act autonomously and contributed to a renewal of a more confrontational leftist politics in the

United States. The legacy of OWS is an emboldened left and a series of campaigns based on the concrete struggles and needs of the ninety-nine percent as exposed on sites like the "We are the 99 percent" Tumblr.[46] One of the most compelling campaigns to emerge out of the OWS WGs was Strike Debt, a movement aimed at politicizing debt resistance and building collective action around health care, housing, and student debt.[47] What is promising about a movement around student debt in particular is that it focuses on mobilizing a specific demographic who, in spite of their many differences, have a clear common interest in relation to issues of educational policy, tuition fees, and youth unemployment. A basis in specific common problems and common interests may lend direct democratic practices a greater functionality than that seen in the OWS camps.

Movements that seek deep and systemic change of capitalist parliamentary democracy should not consider participation to be an unqualified good. The tradition of participatory democracy as articulated in both activism and theory offers a welcome critique of the inadequate nature of representative systems for developing a democratic culture. However, as I have argued throughout this chapter, even under circumstances where participation is intended to offer a radical critique of the inequality that exists within representative politics and capitalist society, the set of values that it invokes often undermines effective democratic power within movements. This set of values that it foregrounds — education, integration, cooperation, listening, self-development, and self-expression — risks undermining the necessarily oppositional and confrontational politics required to dismantle corporate class power, thus preventing movements from politicizing their boundaries for the sake of a desire to embody an inclusive ideal. Participation as a principle within activism should be heavily qualified to reflect more substantive political principles. The organizational principles that participation may indicate — emphases on member autonomy, direct democratic control by the base, and direct action — can work in the direction of building democratic power if they are accompanied by a strong sense of solidarity based on common interests, ideological coherence, shared demands, and a willingness to engage in confrontational tactics. When participation is interpreted in an absolutist fashion or raised up as the central principle of a movement's politics beyond any goals outside of formal decision-making structures, this participatory complex may actually stand in the way of democratic practice.

Notes

The author would like to thank Darin Barney and Jonathan Sterne for their valuable feedback on this essay.

1. The activist critique of neoliberalism began much earlier than this more recent wave of popular protests. The late-1990s "antiglobalization" movement was directly inspired by the 1994 Zapatista uprising against the signing of the North American Free Trade Agreement. Leftist activists adopted the term neoliberalism into their critical lexicon as a way of naming a set of related but distinct practices typical of late capitalism, such as privatization, marketization, deregulation, and financialization, and connecting these practices to increasing inequality, declining public services, the upward redistribution of wealth, precarious working conditions, securitization, and militarization. However, within political movements the term has a directly political function and is not simply a descriptive term. Naming these problems "neoliberalism" allows distinct struggles to be seen as being interrelated and therefore fosters political solidarity across regional, cultural, ethnic, and class differences. The political work of building solidarity against neoliberalism (as opposed to focusing solely on its distinct symptoms or effects in various contexts) has contributed to the recent resurgence of leftist, anticapitalist consciousness and increased public awareness of the historical and ideological nature of the economic transformations that have taken place since at least the 1970s.

2. Occupy Wall Street New York General Assembly, "Principles of Solidarity," released September 23, 2011, http://www.nycga.net/resources/documents/principles-of-solidarity./.

3. Carol Pateman, *Participation and Democratic Theory* (Cambridge: Cambridge University Press, 1970), 17–18.

4. Ibid., 19–20.

5. Ibid., 22–44.

6. Ibid., 42.

7. Ibid., 24.

8. Ibid., 25.

9. Ibid., 27.

10. Ibid.

11. The possibility and meaning of participatory social relations in these social spheres and others are treated in turn in Chris Spannos, ed., *Real Utopia: Participatory Society for the 21st Century* (Oakland, Calif.: AK Press, 2008).

12. Pateman, *Participation,* 34.

13. Jean-Jacques Rousseau, *The Social Contract or Principles of Political Right,* trans. G. D. H. Cole (public domain: 1762), book 2, chap. 11. Quoted in Pateman, *Participation,* 23.

14. C. B. Macpherson, "Democratic Theory: Ontology and Technology," *Democratic Theory: Essays in Retrieval* (Oxford: Clarendon Press, 1973), 24–38.

15. C. B. Macpherson, "Model 2: Developmental Democracy," in *The Life and Times of Liberal Democracy* (Oxford: Oxford University Press, 1977), 44–76.

16. Ibid., 49.

17. Ibid., 71.

18. Ibid., 48.

19. Jodi Dean argues against this reductive equation of communism or socialism with Stalinism in her chapter "Our Soviets," *The Communist Horizon* (London: Verso, 2012), 23–38.

20. David Graeber, *The Democracy Project: A History, A Crisis, A Movement* (New York: Spiegel & Grau, 2013), xix–xx.

21. Occupy released a list of grievances that were mostly aimed at corporate power and inequality. See Occupy Wall Street New York General Assembly, "Declaration of the Occupation of New York City," released September 29, 2011, http://www.nycga.net/resources/documents/declaration/.

22. For a general account of this crisis see chapter 1, "The Disruption," in David Harvey's *The Enigma of Capital and the Crises of Capitalism* (Oxford: Oxford University Press, 2010), 1–39.

23. A large amount of documentation of working group discussions, research, and proposals can be found on the Occupy Wall Street New York General Assembly website on the "documents" page: http://www.nycga.net/documents/.

24. Marina Sitrin, "One No, Many Yeses," in *Occupy! Scenes from Occupied America*, ed. Astra Taylor et al. (London: Verso, 2011), 200.

25. Graeber, "'The Mob Begin to Think and Reason': A Covert History of Democracy," in *The Democracy Project*, 150–207.

26. Here I would note that coercion should not be equated with violence, but that physically occupying space is an assertion of power.

27. Dean, *Horizon*, 210.

28. Dean, "Claiming Division, Naming a Wrong," in *Occupy! Scenes from Occupied America*, ed. Astra Taylor et al., 88.

29. Dean, *Horizon*, 218–19.

30. Ibid., 220.

31. Dean, "Claiming Division," 89–90.

32. Graeber, *The Democracy Project*, xx.

33. L. A. Kauffman, "The Theology of Consensus," in *Occupy!: Scenes from Occupied America*, ed. Astra Taylor et al., 49. It should be noted that this embrace of consensus is not necessarily representative of the majority of those who identify as leftists, but that the consensus position nevertheless exercises a strong moral hegemony within recent movements. Those who pushed for consensus in Occupy were the most active organizers of the initial camp and therefore determined its

decision-making procedures and continued to exert a strong influence on the camps as they developed. Although there appears to have been plenty of criticism and frustration expressed by OWS participants in relation to the consensus process, even those who complained continued to use it. The "Demands" working group, for instance, while upset that their "Jobs for All" proposal with a two-thirds majority still failed to reach the 90 percent support needed for consensus, nevertheless used this process within their working group and continued to engage with the GA. The very fact that within this system such a high degree of support would be required to reject consensus as a decision-making process means that once consensus has been established a very small minority can continue to insist on its use. The only remaining options for those who disagree would be to adapt to this procedure against their better judgment or to become autonomous from the GA by acting without its approval, creating an alternative decision-making body and inviting others to take part, or leaving the movement entirely.

34. Graeber, *The Democracy Project,* 208–70.

35. Keith Gessen, "Laundry Day," in *Occupy! Scenes from Occupied America,* ed. Astra Taylor et al., 200.

36. Graeber, *The Democracy Project,* 203, 217.

37. Occupy Wall Street New York General Assembly, "Principles of Solidarity," adopted September 23, 2011, http://www.nycga.net/resources/documents /principles-of-solidarity/.

38. Graeber, *The Democracy Project,* 220.

39. Ibid., 219.

40. Pateman, *Participation,* 37.

41. Ibid., 40.

42. Graeber, *The Democracy Project,* 227–29.

43. Ibid., 215–17.

44. Mark Greif describes the conflict in his essay "Drumming in Circles," in *Occupy! Scenes from Occupied America,* ed. Astra Taylor et al., 55–62.

45. Gessen, "Laundry Day," 206.

46. "We Are the 99%," http://wearethe99percent.tumblr.com/.

47. In Strike Debt's most well-known campaign, "Rolling Jubilee," the activists raise money to buy others' debt at a reduced rate (as is common practice within the financial industries) and then forgive the debt rather than collect it. So far the campaign has resulted in the canceling of close to 15 million dollars in debt. "Strike Debt," http://strikedebt.org/.

· CHAPTER 3 ·

From *TuniLeaks* to Bassem Youssef

Revolutionary Media in the Arab World

Jillian C. York

Civic media is participatory media—even newspapers and television stations are discovering that they cannot simply deliver information to their audiences. The audience expects to be able to talk back, to share news stories they want to see covered, to offer their interpretation and opinions. Media that doesn't enable participation is likely to be criticized or ignored.

—Ethan Zuckerman, "Civic Media's Challenges and Opportunities,"
speech, Yangon, Myanmar, March 10, 2014

I N TERMS OF PRESS FREEDOM, the Arab world traditionally ranks consistently low.[1] Restrictions on publishing, licensing requirements, and blatant censorship abound throughout the region, leaving countless stories unreported or underreported. Foreign media reporting on the region often comes up against state restrictions, but even when it doesn't, inherent biases sometimes creep in, creating lopsided narratives regarding some of the world's most treasured locales.

The Internet has changed the face of media in the region, allowing disparate communities to connect, activists to seek strength among their allies, and journalists to circumvent many of the restrictions they face. At the same time, the Internet has also been utilized by governments to censor, spy on, and otherwise manipulate their citizens. Three years after the "Arab Spring" uprisings, much has changed, and little has changed. Citizen journalism and new forms of participatory media have emerged, leaving traditional media behind. These new contenders face countless challenges, from raising funds to circumventing censorship, but nevertheless—from *TuniLeaks* to Bassem Youssef—the new face of Arab media is online.

An Emerging Participatory Media

The history of Arab media post-Ottoman rule can be broken down into roughly three eras: a first wave of print and broadcast media, beginning at the turn of the last century and controlled primarily by governments (particularly Egypt's); a second wave, beginning in the 1990s, marked by the emergence of satellite television and the launch of Al Jazeera; and a third wave of participatory media, beginning in the early 2000s with the spread of the Internet throughout the region.[2]

While this third phase was enabled by the expansion of the Internet, it has also been challenged by it. By the year 2000 the Arab world was online, with Syria being the last country in the region to offer Internet access to the public.[3] The eagerness and optimism that allowed the proliferation of the Internet into Arab countries was almost immediately quelled, in some locales, by a desire to control the flow of information. Several governments—including Saudi Arabia, Syria, Kuwait, and Tunisia—quickly found easy and relatively inexpensive ways to block "undesirable" information online, ranging from pornography to opposition politics. Filtering software, purchased from U.S. or European companies, allowed governments to simply input URLs or check off entire prefabricated categories of content that they wished to block.

Such limitations have nevertheless failed to keep the Internet from eternally altering the region's media landscape. Although the methods for controlling information have become more sophisticated, and the stakes of the constant cat-and-mouse game between governments and Internet users have grown, the Internet has demonstrated that the flow of information cannot—will not—be stopped. Despite restrictions and continued conflict, new media continues to flourish in the Arab world.

A New Era

The birth of the blog coincides with the turn of the century. In 1997, American writer Jorn Barger coined the term "weblog" to describe the phenomena of online journaling. The term was later shortened and turned into a verb by Evan Williams, a cofounder of both Blogger (a blog hosting service now owned by Google) and Twitter.[4] The act of blogging was quickly adopted by Silicon Valley elites, followed by pundits and newspapers. In

2006, the *New York Times* launched its first blog, *The Lede;* other newspapers were quick to follow.

Despite low Internet penetration rates throughout much of the Arab world, blogging became popular among the region's cultural elite. By 2006, the number of Arab blogs was cited to be in the range of twenty to twenty-five thousand, with political blogs comprising a small percentage of that number.[5] By 2009, researchers at Harvard University's Berkman Center for Internet and Society estimated that there were close to thirty-five thousand "active" Arabic blogs.[6]

In addition to the large number of Arabic-language blogs, there have, traditionally, been a smaller number of Arab blogs penned in English. While some prominent English-language bloggers from the region have been described as "highly unrepresentative of public opinion in their countries," others have served an international audience well, flagging important movements and highlighting or even translating other bloggers' writing.[7]

In 2007, Marc Lynch observed that in the region, "Blogging remains the activity of a tiny elite, as only a small minority of the already microscopic fraction of Arabs who regularly use the Internet actually write or read blogs," before concluding that, when coupled with crackdowns on bloggers by repressive regimes such as Egypt's or Bahrain's, "it is highly unlikely that blogging will induce wide political change in the Middle East."[8]

As blogs grew in popularity throughout the region, traditional media fought to keep up. Al Jazeera, which launched its television channel in 1996, published its first news site in 2001. In 2003, three years before the launch of its English-language television network, Al Jazeera launched an English-language version of its news site. Rival Al Arabiya followed in 2004 with an Arabic news website, and in 2007 with an English-language one.

The next emergent frontier was social media. With the advent of YouTube, Facebook, and Twitter between 2004 and 2007, activists and would-be pundits now had access to a wealth of platforms where they could express themselves with minimal or no restrictions. In contrast to traditional media—where even opposition publications most often had to be careful not to cross certain lines—these social sites offered a new space in which individuals and groups could challenge the predominant narrative.

New Tactics

Numerous studies have attempted to make sense of the Arab networked public sphere as a whole. For example, a 2009 mapping of the Arab blogosphere found that both "Web 2.0" sites, such as YouTube, and traditional media websites were widely linked to by Arab bloggers.[9] While useful to a point, such generalizations fail to take into account the discrete ways in which networked public spheres have developed in countries across the region.

By the mid-2000s, several governments throughout the region had begun to take note of the ways in which the Internet was being used for political and social organizing. While some—including Tunisia, Bahrain, and Saudi Arabia—had long before wised up to the potential uses of the Internet and censored "subversive" content, others, including Egypt, had left access largely unfettered.

In 2006, the Egyptian government launched a series of crackdowns on prodemocracy activists, arresting numerous members of the growing Kefaya (Enough) movement and bringing charges against a group of outspoken judges.[10] During one such crackdown, a young but prominent blogger named Alaa Abd El Fattah was among those arrested. Fattah's arrest spurred the global blogosphere into action; just a few days after he was detained, the Global Voices Online community launched a campaign to "Google bomb for Alaa," encouraging users to manipulate search engine results to draw further attention to their cause.[11] The "Free Alaa" campaign—which utilized a variety of tactics—resulted in both the intended effect of raising international awareness of Fattah's arrest and, additionally, the unintended effect of creating a new meme: For years since, campaigners throughout the region and beyond have followed the methods and style of the "Free Alaa" campaign.[12]

From 2006 onward, Egypt's online sphere was a hotbed of political activity. Various existing movements, including April 6 and Kefaya, took to online organizing, while others were born from the flurry of online activity. One example of the latter was a public movement against torture that utilized new tools such as YouTube and Flickr to document police brutality. One infamous video, shot in 2006 and posted by Wael Abbas, depicted a bus driver screaming for mercy as police officers sodomized him with a rod. The officers reportedly sent copies of the video to the driver's coworkers to humiliate him.[13]

The use of intermediary corporate platforms for such purposes unearthed a new problem, however; in 2007, antitorture activist Abbas found that his YouTube account had been suspended after he posted videos containing "images of torture, police brutality, demonstrations, strikes, sit-ins and election irregularities."[14] Similarly, in 2011 (after the uprising), prominent political activist and blogger Hossam Hamalawy found content removed from Flickr's photo-hosting service after he posted images of police officers obtained during an activist raid on Egypt's state security offices.[15] These acts of corporate censorship—predicated on each company's own set of rules—continue to raise questions for activists and bloggers around the world.

The Tunisian online public sphere also experienced a spurt of development in the mid-2000s with the birth of *Nawaat,* an opposition publication founded in 2004 by Tunisian exiles in Europe. The positioning of *Nawaat*'s founders meant that they were able, with relative safety, to publish articles challenging the regime of Zine El Abidine Ben Ali and its human rights abuses. The adversarial site wasn't accessible for long inside Tunisia, however; just a few months after its launch, the site was blocked, prompting its editors to write: "We will not yield to police blackmail. We will continue our fight against the Tunisian dictatorship, with the same perseverance and a spirit of renewal more creative than ever."[16]

Nawaat soon found itself at the cutting edge of digital activism when, in 2007, it embedded videos from YouTube (which was blocked in the country) into Google Earth, which was accessible, allowing Tunisians to see documentation of human rights abuses and torture occurring within their country. This method of circumventing censorship allowed Tunisians access to such material for perhaps the first time.[17]

Diasporic Connectivity

As adversarial journalism became a common experience for young Internet users in Egypt and Tunisia, Palestinian users throughout the Middle East seized upon another opportunity: connectivity with their far-flung relatives. For decades, Palestinians in the Middle East needed to find creative ways of establishing contact due to obstacles placed on them by the respective governments under which they lived; for example, a Palestinian resident of Israel could not make a telephone call to a relative in Beirut because calls between the two countries were blocked.

Between 2000 and 2009, Internet usage expanded within the Palestinian territories, increasing its reach from 1 to more than 14 percent of the population. Meanwhile, Internet use was also growing in Palestinian refugee camps in Lebanon, Jordan, and Syria, allowing young Palestinians to connect to one another. Researcher Miriyam Aouragh has posited that the Internet has become "a mediating space through which the Palestinian nation is globally 'imagined' and shaped," allowing the "fragmented Palestinian nation" to begin to "reunite."[18]

This connectivity has also allowed Palestinians in the territories, refugee camps, and the diaspora to focus on a set of common advocacy tactics, including the movement for boycott, divestment, and sanctions (BDS) against Israel first called for by activists in Palestine in 2005. The movement for BDS has relied heavily on the Internet for communications between Palestinians and activists in the diaspora and those on the ground, as well as for the purpose of organizing protests and other events worldwide.

The Internet has also enabled reporting on Palestine that was previously impossible. *Electronic Intifada*, "an independent online news publication and educational resource focusing on Palestine, its people, politics, culture and place in the world,"[19] was founded in 2001 during the Second Intifada, an uprising against Israeli occupation. With reporters both on the ground in Palestine and throughout the diaspora, *Electronic Intifada* is one of the few publications in English focused on Palestinian news, and thus offers a unique glimpse into a country where reporting from both inside and outside is often flawed.

"Courage Is Contagious"

Prior to the Arab Spring, the last time an Arab publication had been involved in a major act of whistle-blowing was in 1986, when the Lebanese paper *Al-Shiraa* exposed the Iran–Contra affair. But in early 2011, not long after *WikiLeaks*' release of diplomatic cables leaked by Chelsea Manning, Al Jazeera embarked on a whistle-blowing project of its own, releasing a cache of sixteen hundred documents from more than a decade of Israeli–Palestinian negotiations. The release was quickly overshadowed by the nascent uprising in Egypt, but nevertheless helped renew the possibilities of whistle-blowing in the region.

Though it went largely unnoticed outside the country at first, another publication had also just become involved in leaking. *Nawaat*, the adversarial Tunisian publication, had quietly struck a deal with *WikiLeaks* that allowed it first access to diplomatic cables relating to Tunisian affairs, which they published on a separate—and officially unassociated—site called *TuniLeaks*. The leaks were contextualized and translated into French, with the aim of reaching a wide Tunisian audience.

The leaked cables made plain the Ben Ali regime's corruption and depravity, and also demonstrated the United States' knowledge of human rights abuses in the country. And the timing of their release was no coincidence; as *Nawaat* cofounder Sami Ben Gharbia would later attest, the cables "helped tip the balance" toward revolution.[20]

The success of *TuniLeaks* inspired another offshoot, this time in neighboring Morocco. Emulating its source of inspiration, *MoroLeaks* sought to contextualize and translate cables related to Moroccan corruption and human rights abuses. Influenced by events in Tunisia and Egypt and helped along by online calls for protest, in February 2011 demonstrations sprang up throughout the country. Before long, however, sly reforms from the Moroccan government were coupled with sometimes violent crackdowns against protesters to quash the growing movement. Nevertheless, *MoroLeaks,* and *WikiLeaks* more broadly, proved an influence on Moroccans online; as one blogger wrote, "The leaked documents have transformed the notions of sovereignty and borders. They have proved one new and undeniable reality: information is queen. How to master this information and turn it into one's favor remains to be seen."[21]

A selection of cables were also later translated and contextualized by the Lebanese paper *Al Akhbar* and the Egyptian paper *Al Masry Al Youm,* both of which were among *WikiLeaks'* more than seventy global publishing partners.[22] This second round of media partners was selected by *WikiLeaks* after an initial round of partnerships resulted in the publication of a mere 2 percent of the possessed cables.[23]

Throughout the rest of the region, citizens read, dissected, and blogged about the cables. Amidst several governments' attempts to censor access to *WikiLeaks* material, polls demonstrated strong support for the whistle-blowing platform: A regional survey conducted by the Doha Debates found that "six out of ten Arabs believe the world is better off with *WikiLeaks*" and that "nearly three quarters [of those polled] would like to see the whistle-blowing website publish more on the Arab world."[24]

Julian Assange, the editor in chief of *WikiLeaks* who has said he lives by the motto that "courage is contagious,"[25] spoke of the influence the cables had on the Arab uprisings of 2011: "Just before the Berlin Wall fell, everyone thought that it was impossible. It wasn't that people suddenly received a lot of new information. Rather, the information that they received was that a large majority of people had the same beliefs that they had, and people became sure of that, and then you have a sudden switch, a sudden state change, and then you have a revolution. So, people becoming aware of what their beliefs are, what each other's beliefs are, is something that introduces that truly democratic shift."[26] Assange's perspective, shared by other scholars such as Zeynep Tufekci,[27] emphasizes the need for preexisting belief structures in information activism. In other words, it is not merely enough to make information available or transparent; the political impetus must already be in place.

Participatory Media in Postrevolutionary Egypt: A Case Study

After the dust settled from the uprisings which swept across the Arab world in 2011, one thing was clear: Telling the stories of a complex region would require better, more participatory local media. While mass media inside the region has too often been encumbered by restrictions on the press, foreign media all too frequently presents the diverse region as a single entity, backward in practices and oppressed by religious forces and conflicts. The foreign media's oversimplified characterization of the Internet's role in the so-called Arab Spring illustrates this point.

In Egypt—often referred to by locals as "umm al-dunya" or "mother of the earth" and historically the Arab world's cultural compass—awareness of the need for new media norms was entrenched among the country's online community. The January 25 uprising—which initially relied on social media to coordinate and spread the word about protests—inspired some of the country's most powerful media developments in decades.

One example is Mosireen, a media collective that grew out of the early days of protest in Cairo's Tahrir Square. Filmmakers and other citizen journalists congregated at a media tent in the square and began to build an archive of footage shot during the revolution. After the last of the encampments were cleared, the collective found a space and began offering workshops and a gathering point for filmmakers. Realizing that their videos could only spread as far as the country's Internet access allowed (to roughly 40 percent of the population in 2011),[28] the collective set up Cinema Tahrir, a series of outdoor screenings of footage filmed during the revolution.[29]

Another example of the impact of video content in Egypt comes via Bassem Youssef, a heart surgeon turned television satirist. In March 2011, just a month after Egyptians ousted former president Hosni Mubarak from his autocratic rule, a little-known cardiac surgeon emerged on YouTube, challenging the Egyptian state media's attempts to spread conspiracy theories about Tahrir Square's protesters.[30] Within a month the surgeon had the most-viewed YouTube channel in Egypt and was being hailed in the American media as "Egypt's Jon Stewart."[31] Soon he had his own satirical television program, styled after Stewart's, and he would go on to be named by *Time* magazine as one of the one hundred most influential people in the world.[32]

"We did all of this for the sake of a different media, a media that respected people's minds and intelligence and at the same time keeping them informed and helping them to combat lies and misinformation that plagued many of the media outlets," said Youssef during a 2014 speech at the Deutsche Welle Global Media Forum. "As we did that, we never claimed to be freedom fighters or political activists. We believed that it was enough to be the normal everyday people who did not buy into the everyday life of propaganda that is full of lies and deceit."[33]

Notably, while Bassem Youssef's program, *El Bernameg,* moved from YouTube to satellite television, it continued to appear in clip form on YouTube. "Multi-platform productions are the new business model," writes Jeffrey Ghannam of the phenomenon.[34] Ghannam further states: "The blend of new and traditional media is likely to bring new voices and multi-platform offerings to audiences."[35]

There is a famous saying about literature in the Middle East that goes: "Cairo writes, Beirut publishes, Baghdad reads." Yet, when it comes to the news, Cairo seems to be the one doing the reading: Daily newspaper circulation in the country stands at more than 4.3 million, the highest in the Arab world.[36] Despite that, independent media has struggled post-January 25.

Egypt's latest wave of media is not, inherently, participatory media; however, its inclusiveness ensures that a broader swath of Egyptian voices are heard in its reporting. Attempts to integrate social media into reporting—as well as a greater effort by young journalists to engage with the public through social media—have been successful, contributing to a shift in perspective.

In 2013, the publisher of Egypt's popular newspaper *Al Masry Al Youm* shut down its English-language news subsidiary, *Egypt Independent,* after a successful but short two-year run. The small paper, known for its progressive outlook and strong cultural writing,[37] had effectively served a niche: catering to both English-speaking and expatriate Egyptians, as well as foreign readers interested in keeping up with the left side of Egyptian politics.

Out of the closure, which included the firing of twenty-five staff, a new publication was born. *Mada Masr* was launched in mid-2013 by a team largely comprised of former *Egypt Independent* staff, including its editor in chief, Lina Attalah. A young but veteran journalist, Attalah has spoken honestly about the challenges of producing the publication in Egypt's current context. In a 2013 interview, she said:

> The Egyptian media establishment is growing increasingly limiting to independent media practice and practitioners. They have been doing so through a series of policies and directions that haven't come out of research or even the desire to find new robust solutions in response to growing economic challenges. There is only a process of reproducing these challenges. For example, Egyptian media prints in order to make money, because web advertising doesn't pay the bills, it's actually the print ads that do. That's why [*Al Masry Al Youm*] still keeps printing, and that's why [*Egypt Independent*] started printing. And then in defending itself, the institution uses a relic of the practice they let go of. So, to the world, they say they ended the print edition of EI and kept the website in line with the world going digital, when in reality they downsized the whole operation and let go of the team that produces content for both the website and the print edition. We have initially said that we should invest in the web as much as we can rather than divert and go print. But they never listened.[38]

Despite facing many challenges, *Mada Masr* has succeeded in gaining the attention of Egyptian readers as well as international journalists; the publication's writers are regularly quoted in prominent foreign media. *Mada Masr* has taken a stand, publishing pieces that might be considered adversarial to all parties in the political sphere, as well as articles from prominent activists inside Egypt's prisons, who might be considered personae non gratae by other publications.

Although considering the Egyptian media sphere provides only one example of how participatory media affected the revolutionary era, it is an example that may serve others in the region as a vision of what is possible, just as *WikiLeaks* did for *TuniLeaks* and *TuniLeaks* did for *MoroLeaks*. Despite the vast differences between countries in the region, successes—particularly in the revolutionary context—often have a ripple effect among the digital elite.

A Closing Window?

In the early days of 2011, many in the Arab world felt that anything was possible. Tunisia and Egypt had just overthrown dictators who had ruled for decades, and several other countries seemed poised to do the same. But by the end of the year, as violent conflict steadily displaced nonviolent protests in Syria, Bahrain, and elsewhere, much of that early hope was lost.

As conflicts grew and Syria slipped into civil war, many of those optimistic about the Internet's potential to promote change became increasingly aware that a second battle was brewing—this one between Internet users and their governments. Although the uprising of 2011 brought an end to Internet controls in Tunisia, in many other countries throughout the region attempts to control information online were only beginning. In 2012, both Lebanon and Iraq introduced draconian bills to regulate online content.[39] Jordan followed in 2013 with an amendment to the Press and Publications Bill that effectively required online news providers to obtain licenses or have their sites blocked.[40] Qatar has made a similar attempt at regulation with a cybercrime bill that inflicts harsh punishments on those who "[infringe] on the social principles or values" of the nation.[41]

In 2012, emerging news reports demonstrated what many activists in the region had long suspected: Western companies were selling sophisticated surveillance tools to Arab governments.[42] Additional research by the University of Toronto's Citizen Lab has since shown the global reach of such tools.[43] Coinciding with the spread of online censorship and other information controls—including legal restrictions and the detention of bloggers and online activists—this development has led many to conclude that governments may finally be winning the cat-and-mouse game.

Still, those who are crafty enough continue to find ways around restrictions. The Jordanian media collective *7iber* was blocked after the enactment of the Press and Publications Law, but has found ways to circumvent the censorship by posting articles in various places, including Facebook (which remains accessible).[44] Egyptian publications such as *Mada Masr* have contended with the arrests of their comrades, but have struck back by publishing their letters from inside prison, raising the profiles of the imprisoned and stoking anger among the populace.

At the same time, from inside countries like Syria and Iraq, there is scant good news. At the time of this writing, both countries are under attack from an extremist group known as Islamic State (IS), which seeks to set up a caliphate across the Levant. The networked public spheres that had begun to emerge in both countries in the mid-2000s have been broken down, their participants in exile, in battle, or dead. On the other hand, IS has taken to social media, posting updates on its advances as well as images of beheadings and other brutality.[45]

The future of the media in the Arab world is dependent on a number of factors, few of which seem within the control of its creators. Yet, with the rise of participatory formats over the past decade, it seems impossible to go back to the days of staid, state-controlled print media. As Internet penetration continues to increase throughout most of the region, new forms of interaction online will also proliferate, opening up vast possibilities for change.

The images and ideas afforded to the general public thanks to participatory media are certain to have an impact on societies. What was once censored or otherwise unavailable is now widespread, from calls for revolution to images of beheadings. Even amid the onslaught of increased online censorship and surveillance throughout the Arab world, the use of social media—for reporting, disseminating information, and movement building as well as "regular" communication—continues.

As Ethan Zuckerman's "Cute Cat Theory" suggests, Internet tools designed for ordinary consumers are often used by activists because they are difficult for governments to censor without also censoring innocuous content (such as cute cat videos on YouTube), an inevitability they seek to avoid, since the disappearance of inoffensive content can alert nonactivists to government censorship.[46] Indeed, throughout the Arab world— both prior to and particularly following the events of 2011—this theory has proven demonstrably true, as activists share political content that is then reshared and commented on by their friends and communities. The effects of this in the Arab world are palpable; from the aforementioned examples as well as many others, it is clear that participatory media can be a powerful tool to harness energy and attention toward a diverse set of causes, in spite of government attempts to control information and infiltrate networks. At the same time, as usage grows among a populace, competition for attention grows as well, making it more difficult for activists to reach beyond a core audience. The revolution may be tweeted, but Twitter alone will not spark revolution.

Notes

1. Reporters without Borders, "World Press Freedom Index 2014," http://rsf.org/index2014/en-index2014.php/.

2. Dr. Sahar Khamis and Katherine Vaughn, "Cyberactivism in the Egyptian Revolution: How Civic Engagement and Citizen Journalism Tilted the Balance," *Arab Media and Society* 14 (2011), http://www.arabmediasociety.com/index.php?article=769&printarticle/.

3. Hasna Askhita, "The Internet in Syria," *Online Information Review* 24, no. 2 (2000): 144–49.

4. Jenna Wortham, "After 10 Years of Blogs, the Future's Brighter Than Ever," *Wired,* December 17, 2007, http://www.wired.com/entertainment/theweb/news/2007/12/blog_anniversary/.

5. Marc Lynch, "Blogging the New Arab Public," *Arab Media & Society* 1 (2007), http://www.arabmediasociety.com/?article=10/.

6. Bruce Etling et al., "Mapping the Arabic Blogosphere: Politics, Culture and Dissent," *New Media & Society* 12, no. 8 (2010): 1225–43.

7. Marc Lynch, "Blogging the New Arab Public."

8. Ibid.

9. Bruce Etling et al., "Mapping the Arabic Blogosphere: Politics, Culture, and Dissent," Berkman Center Research Publication No. 2009-06, June 2009, http://cyber.law.harvard.edu/sites/cyber.law.harvard.edu/files/Mapping_the_Arabic_Blogosphere_0.pdf/.

10. Rory McCarthy, "Cairo Clamps Down on Dissent," *Guardian,* May 8, 2006, http://www.theguardian.com/world/2006/may/08/worlddispatch.egypt/.

11. Haitham Sabbah, "Google-Bombing for Alaa: Press Release," *Sabbah Report,* May 10, 2006, http://sabbah.biz/mt/archives/2006/05/10/google-bombing-for-alaa-press-release/.

12. Jillian C. York, "The Arab Digital Vanguard: How a Decade of Blogging Contributed to a Year of Revolution," *Georgetown Journal of International Affairs* 13, no. 1 (2012).

13. Sarah O. Wali, "Man Who Posted Videos of Police Torture and Rape Hides from Mubarak Regime," *ABC News*, February 3, 2011, http://abcnews.go.com /Blotter/egypt-police-brutality-documented-blogger-wael-abbas-now/story ?id=12831672/.

14. CNN, "YouTube Shuts Down Egyptian Anti-Torture Activist's Account," *CNN*, November 29, 2007, http://www.cnn.com/2007/WORLD/meast/11/29 /youtube.activist/.

15. Jennifer Preston, "Ethical Quandary for Social Sites," *New York Times*, March 27, 2011, http://www.nytimes.com/2011/03/28/business/media/28social .html?pagewanted=all&_r=1&/.

16. *Nawaat*, "Nawaat.org Suspendu. Nawaat.org Survivra!" November 29, 2004, http://nawaat.org/portail/2004/11/29/nawaatorg-suspendu-nawaatorg -survivra/.

17. *Nawaat*, "Google Earth Bombing for a Free Tunisia," May 28, 2008, http:// nawaat.org/portail/2008/05/28/google-earth-bombing-for-a-free-tunisia/.

18. Miriyam Aouragh, Palestine *Online: Transnationalism, the Internet and Construction of Identity* (London: I. B. Taurus, 2006), 4–10.

19. Electronic Intifada, "About the Electronic Intifada," http://electronic intifada.net/about-ei/.

20. Yasmine Ryan, "Breaking through Information Monopoly," *Al Jazeera English*, October 6, 2011, http://www.aljazeera.com/indepth/features/2011/10 /2011104115312389414.html/.

21. Hind Soubaï Idrissi, "Wikileaks or the New Frontline," *Talk Morocco*, December 28, 2010, http://www.talkmorocco.net/articles/2010/12/wikileaks-or -the-new-frontline/.

22. Lisa Lynch, "The Leak Heard Round the World? Cablegate in the Evolving Media Landscape," in *Beyond WikiLeaks: Implications for the Future of Communications, Journalism and Society,* ed. Benedetta Brevini et al. (London: Palgrave Macmillan, 2013), 56–57.

23. Lisa Lynch, "The Leak Heard Round the World?" 57.

24. *The Peninsula*, "Majority in Arab World Backs Wikileaks," February 16, 2011, http://thepeninsulaqatar.com/news/qatar/142783/majority-in-arab-world -backs-wikileaks/.

25. *Al Jazeera English,* "Julian Assange Interview," December 22, 2010, http://www.aljazeera.com/programmes/frostovertheworld/2010/12/20101222 8384924314.html/.

26. Amy Goodman, "Julian Assange of WikiLeaks & Philosopher Slavoj Žižek in Conversation with Amy Goodman," *Democracy Now,* transcript, July 5, 2011,

http://www.democracynow.org/blog/2011/7/5/watch_full_video_of_wikileaks
_julian_assange_philosopher_slavoj_iek_with_amy_goodman/.

27. Zeynep Tufekci, "New Media and the People-Powered Uprisings," *Technology Review,* August 30, 2011, http://www.technologyreview.com/view/425280 /new-media-and-the-people-powered-uprisings/.

28. "International Telecommunications Union," https://www.itu.int/net4/itu -d/icteye/.

29. Bel Trew, "Egyptian Citizen Journalism 'Mosireen' Tops YouTube," *Ahram,* January 20, 2012, http://english.ahram.org.eg/News/32185.aspx/.

30. Soraya Morayef, "Cairo 360 Presents: The Bassem Youssef Show (B+)," *Cairo 360,* March 23, 2011, http://www.cairo360.com/article/tvanddvd/1619/cairo -360-presents-the-bassem-youssef-show-b/.

31. Robert Siegel, "Egypt Finds Its Own 'Jon Stewart,'" *NPR,* April 15, 2011, http://www.npr.org/templates/transcript/transcript.php?storyId=135444956/.

32. Jon Stewart, "TIME 100: Bassem Youssef," *Time,* April 18, 2013, http:// time100.time.com/2013/04/18/time-100/slide/bassem-youssef/.

33. Bassem Youssef, "An Interview with Bassem Youssef," *Deutsche Welle Global Media Forum.* http://www.dw.de/special-guest-bassem-youssef/av -17747241/.

34. Jeffrey Ghannam, "Digital Media in the Arab World One Year after the Revolutions," *CIMA/NED,* March 28, 2012, http://cima.ned.org/publications /digital-media-arab-world-one-year-after-revolutions/.

35. Ibid.

36. Romesh Ratnesar, "Egypt: Not Just the Facebook Revolution," *Bloomberg Businessweek,* June 2, 2011, http://www.businessweek.com/magazine/content/11 _24/b4232062179152.htm#p1/.

37. Ibid.

38. Mia Jankowicz, "Interview with Lina Attalah," *ArtTerritories,* http://www .artterritories.net/?page_id=3308/.

39. Jillian York, "Proposed Laws in Lebanon and Iraq Threaten Online Speech," Electronic Frontier Foundation, April 2, 2012, https://www.eff.org/deeplinks /2012/03/proposed-laws-lebanon-iraq-threaten-online-speech/.

40. Jillian York, "Jordan Takes a Disappointing Turn toward Censorship," Electronic Frontier Foundation, June 3, 2013, https://www.eff.org/deeplinks/2013/06 /jordan-takes-disappointing-turn-toward-censorship/.

41. *Doha News,* "Qatar's New Draft Cybercrime Law Raises Concerns over Freedom of Online Expression," May 30, 2013, http://dohanews.co/qatar-new -draft-cybercrime-law-raises-concerns-over/#ixzz2UlCXwWD9/.

42. Vernon Silver, "Spyware Leaves Trail to Beaten Activist through Microsoft Flaw," *Bloomberg,* October 10, 2012, http://www.bloomberg.com/news/2012–10–10 /spyware-leaves-trail-to-beaten-activist-through-microsoft-flaw.html/.

43. Morgan Marquis-Boire et al., "You Only Click Twice: FinFisher's Global Proliferation," *CitizenLab,* March 13, 2013, https://citizenlab.org/2013/03/you-only -click-twice-finfishers-global-proliferation-2/.

44. Lina Ejeilat, "Jordanian Government Blocks 7iber Again," *7iber,* July 1, 2014, http://7iber.net/2014/07/jordanian-government-blocks-7iber-again/.

45. Alice Speri, "ISIS Fighters and Their Friends Are Total Social Media Pros," *VICE News,* June 17, 2014, https://news.vice.com/article/isis-fighters-and-their -friends-are-total-social-media-pros/.

46. Ethan Zuckerman, "Cute Cats to the Rescue? Participatory Media and Political Expression," in *Youth, New Media and Political Participation* (Cambridge, Mass.: MIT Press, forthcoming), http://hdl.handle.net/1721.1/78899/.

CHAPTER 4

Think Outside the Boss

Cooperative Alternatives for the Post-Internet Age

Trebor Scholz

THE PAST DECADE has not only been about the sharing economy— the Handys, Feastlys, TaskRabbits, and Ubers; it has also been marked by the complete inability of traditional unions to answer to the new demands of a dispersed workforce, sharp attacks on employment and worker rights, faster PCs, Big Data, cloud computing, and the crash of the financial system in 2008. At the same time, automation started to affect an ever-expanding number of professions—from doctors, lawyers, and professors, to cooks and farmers.

Over the past five years, future-of-work research groups have popped up, not only in universities and nonprofits but also in corporate research centers. The definition of "digital labor" doesn't only describe value capture on the social web (Facebook, Google, Microsoft, etc.); it also includes wage labor in the crowdsourcing industry. And the definition of digital labor has to go beyond "the immaterial"; it has to account for the largely invisible domestic workforce and the supply chains: the hundreds of thousands of Foxconn workers, for example, who assemble the Xbox and Apple products in China.

At the recent Digital Labor conference at The New School, my colleague McKenzie Wark proposed that the modes of production we appear to be entering are not quite capitalistic as classically described.[1] "This is not capitalism," he suggested. "This is something worse." Indeed, it is likely that we will look back on this era and understand it as a turning point for both lifestyles and the nature of work.

From freshman to senior, university students are warned about the precariousness of their future work lives as if they were boarding a self-driving Google car headed straight for Armageddon. But this "crisis" is not an inevitable process; there isn't just one future of work. Robotic abundance, layoffs, and wage stagnation do not just appear out of the blue. They are developments that are orchestrated and that can be resisted. The realities of twenty-first century labor may be somber, but workers are not despondent.

Let us apply the power of our social and technological imagination to build forms of cooperation and collaboration.

Three years ago, who would have predicted that IKEA, Walmart, and Amazon would be hit by waves of strikes and walkouts?[2] Or that New York City would introduce paid sick leave and the city's taxi drivers would form the NYC Taxi Association?[3] Now there is also a driver-owned ride rental service called La'Zooz and a co-op-based version of eBay in Germany, named Fairmondo.[4] Who would have believed that in May 2014, fast-food workers from New York City to Mumbai, Paris, and Tokyo would coordinate a global strike, picketing McDonald's, Burger King, and Pizza Hut, fighting just-in-time scheduling and demanding a fifteen-dollar minimum wage floor and benefits?[5]

Who, in the face of the Uberization of everything from haircuts to medical services, stands in solidarity with the poorest, most exploited workers? The digital labor movement is taking off. For hackers, "long tail workers," and labor activists, now is the time to form (apps-based) worker-owned cooperatives. I'm calling it *platform cooperativism,* and examples of it already exist. This is the time to form or join innovative unions and support design interventions that allow for moments of solidarity among geographically dispersed digital workers.

From April to June 2013, Bank of America, Citigroup, Wells Fargo, JPMorgan Chase, Goldman Sachs, and Morgan Stanley had their highest-ever quarterly profits: a combined 42.2 billion dollars.[6] Already three decades before that, in the late 1970s, the wages of American workers had started to stagnate while their productivity consistently increased, leading to today's situation, where almost half of all Americans are economically insecure: They cannot afford basic needs like housing, food, or health care.[7] Without income disparity and wealth inequality, companies like Handy or Uber—and the "sharing economy," more broadly—would falter.

Currently, 42.6 million Americans—the unpaid interns, the domestic workers, the freelancers, the university adjuncts and schoolteachers, the precarious culture workers, the crowd workers, and the independent contractors—are working in the contingent sector of the economy, getting by without one single income from one job alone. They are functioning as creative, so-called microentrepreneurs—called upon to be warriors of their own creativity, and occupying a social position characterized by insecurity and instability. While their tax status identifies them as small business owners, they are in fact microearners and not entrepreneurs. Under the banner of coolness, choice, flexibility, and self-reliance, market pressures are passed on to these workers.

Crowdsourcing, the process of obtaining content, services, or ideas by soliciting contributions from a large group of people, is emblematic of this trend. Labor companies such as Amazon, Handy, and Uber are hiding behind the mask of the tech startup, preying on workers' vulnerabilities at a distance.

Amazon.com

Amazon Mechanical Turk, also called AMT or MTurk for short, is a *public* online crowdsourcing system that was founded by Amazon.com in November 2005. Based on the so-called Agreva technology that Amazon acquired, this de facto Internet labor brokerage is designed for corporate labor management. MTurk is based on the idea that certain tasks are easy to perform for the human "turking class," but difficult or impossible to execute for computers. While people are powering the system, it is meant to feel like a machine to users; humans are seamlessly embedded in the algorithm. Companies using MTurk tend to ignore the fact that it is human beings and not algorithms who are toiling for them—people with very real human needs and desires. Amazon founder and CEO Jeff Bezos makes his pet project sound quite straightforward and harmonious. MTurk, he writes, is a marketplace where "folks who have work meet up with folks who seek work." But then it all sounds far less mellifluous when one considers that, when faced with labor conflicts, Bezos's company remains strictly hands-off, insisting that they are merely providing a technical platform.

Consider the example of Stephanie Costello, a fifty-year-old living at the edge of a desert town in the American Southwest.[8] Despite earning an associate's degree in nursing, she has been unable to find work. Moshe Marvit relays the story of how in 2007, during slow periods at a dull office job, she started to fill in online surveys for Amazon Mechanical Turk.

A year later, when the financial crash hit hard, she lost the job and took to "turking" full-time. Today, 18 percent of turkers report that they are sometimes or always reliant on MTurk to "make basic ends meet"—a seemingly futile endeavor, given that the average wage for inexperienced, novice workers fluctuates between two and three dollars per hour.[9] In 2013, Costello worked roughly sixty hours a week for MTurk, searching for and performing so-called human intelligence tasks (HITs). Costello could make 150 dollars one week, only to take home a mere 50 dollars the next.

Prospective "quasi-employers"—companies, researchers, and individuals in need of a variety of people to perform particular tasks—place the tasks on the site. Workers earn their wage by performing a large number of these tasks. Strikingly, Amazon absolves itself of any and all responsibility related to conflicts between workers and these quasi-employers. If workers are unsatisfied, their gripe is with the quasi-employers, not with Amazon. But with the need to perform tasks quickly to accrue a living wage, when could Costello find the time to organize? The lack of time becomes an instrument of oppression.

The size of the turker workforce is a known unknown. Amazon claims that it has 500,000 registered workers, but the 2011 paper "Amazon Mechanical Turk: Goldmine or Coal Mine?" estimates that the number of active workers is between 15,059 and 42,912. But within this group, it is only 3,011 to 8,582 workers who perform 80 percent of all HITs.[10] Today, the number might be even lower because international workers can no longer register for MTurk. Amazon does not release updated statistics.

When looking at the digital labor landscape overall, it is not always clear who the workers are and where they are located. It is difficult to determine how much workers actually earn and how many hours they put in. What are their motivations? How many workers actually toil on a given site on a daily basis and how many have merely created an account and rarely returned? Are people committed to only one company? Or do many split their days between UberX, TaskRabbit, and perhaps 99designs?

At the Digital Labor conference in 2014 at The New School, Karen Gregory, lecturer at the Center for Worker Education at the City College of New York, said, "Solidarity without critical race theory is scary."[11] Her statement becomes especially urgent when investigating the global supply chains for crowdsourcing; we find that the headquarters of companies such as oDesk are mostly located in overdeveloped countries, while economically developing countries are often the sites where underpaid workers reside. To ignore such racial/colonial dynamics would indeed be sinister.

An emerging digital labor movement will have to answer these questions. A study by the American software company Intuit suggested that by 2020 independent contractors and freelancers will constitute 40 percent of the American workforce. Digital laborers are an ever-growing subset of this population. In the wake of its merger with eLance, the online labor broker oDesk now claims a standing reserve of eight million workers. A report by http://crowdsourcing.org found that there are more than six million crowd workers, with roughly two million living in the United States.[12]

Silicon Valley wealth has never been so concentrated. A handful of young, rich companies like Amazon become dangerous when they feel threatened by shareholder dismay. While year after year Amazon makes a show of its zero-profit model, it fails to acknowledge that the company is, in fact, highly profitable: It simply opts to invest billions of dollars into research and development—an investment in the company's future—rather than paying shareholders.[13] Such practices should be understood as a creative experiment in holding down wages. Amazon can exercise its power, but it also shows a lack of ethics. Strikes in its warehouses, atrocious negotiations with publishers, and below-minimum wage payments for MTurk workers form only the top of this disagreeable iceberg.

Importantly, however, Amazon is not the sole perpetrator; the entire crowdsourcing industry operates in a legal gray zone, characterized by deregulation and lacking enforcement of legally guaranteed labor standards. The average hourly rate at CrowdFlower is two to three dollars, for example, and while one might think that the U.S. Department of Labor should enforce labor standards and punish such violations, they are so underfunded that it would take the department about two hundred years to follow up any serious complaints regarding Fair Labor Standards Act (FLSA) violations.

Nevertheless, several workers have filed class-action lawsuits against crowdsourcing companies, and even a single judge ruling in favor of the workers could impact the entire crowdsourcing industry. In October 2012, CrowdFlower worker Christopher Otey sued the company, arguing that it failed to pay FLSA-specified minimum wages—$7.25 per hour at the time—to its American workforce. At this writing, it looks hopeful for Otey.

Differences

Despite these developments, every morning, before you can pour the next cup of coffee, an app like Lyft or Sweetch muscles in on yet another industry, potentially making a traditional job obsolete faster than you can say "Amazon.com." This is not the world of Henry Ford, for whom it wasn't only important to make production processes more effective: He at least understood his workers as consumers of the products they were making.

Ford knew that even hectomillionaires can only buy so much; they cannot spend enough of their wealth to have a truly significant impact on the economy.

A close associate of Henry Ford observed that "cars are the byproducts of his real business, which is the making of men."[14] Today's executives, especially in the digital labor surveillance complex, have short-term profits on their minds; the well-being of "providers"—formerly known as "workers"—is irrelevant. You can, for example, sit in on conferences about the "sharing economy" for days without ever hearing any concern for their situation.

Speaking to young tech entrepreneurs in 2010, Lukas Biewald, the CEO of CrowdFlower, shared that "Before the Internet, it would be really difficult to find someone, sit them down for 10 minutes and get them to work for you, and then fire them after those 10 minutes. But with technology, you can actually find them, pay them the tiny amount of money, and then get rid of them after you don't need them anymore."[15]

To further illustrate the disregard for workers, you can also follow the linguistic trapeze acts of a company like TaskRabbit, which describes its workers as "rabbits." TaskRabbit joins MTurk and other companies, attempting to make workers forget that they are workers while at the same time perhaps conveying a brutal truth when tacitly comparing them to animals.

On the other hand, you might know crowd workers who are doing just fine. Indeed, for the highly skilled, freelancing can be advantageous in the overdeveloped world, and even in the economic developing world; as Microsoft Research has concluded, "Paid crowdsourcing has the potential to improve earnings and livelihoods in poor communities around the world."[16]

Indeed, other models exist: Crowdsourcing companies like Mobile-Works pay the minimum wage specified by a worker's country of residence.[17] While such a model has its own difficulties, MobileWorks shows that the crowdsourcing industry would not go to rack and ruin if adherence to minimum wage standards were introduced. The search for other ethical crowd work upstarts is on!

While there are definitely advantages to being an "independent," it also signals the loss of legal protections with regard to minimum wage, overtime, workplace harassment, paid overtime, employer-financed health insurance, and the eight-hour workday. And this is the same eight-hour workday that organized labor had fought over for at least one hundred years. Under the New Deal, these rights were explicitly reserved only for employees, not for all citizens. In poorer countries, increased reliance on crowdsourcing companies headquartered in Silicon Valley would also reinforce colonial relationships that exploit cheap labor without building up infrastructures that would, systemically and for the long term, allow the workers to improve their situation.

What we need today are employee-like rights for all. One way of reaching this objective would be to loosen the definition of employment to include the work realities of everybody in the global population.

The End Game

In *Who Owns the Future,* Jaron Lanier suggests that while the Internet is poised to rid American society of its middle class, micropayments could become its savior.

In the 1990s, right-wing acolyte Newt Gingrich welcomed the upward spiral of the Internet as a way to "empower elites and reevaluate forms of government."[18] In his latest book, *Average Is Over,* conservative economist Tyler Cowen proposes that there will soon be an American superclass—a "hyper meritocracy" composed of somewhere between 10 percent and 15 percent of the population—of individuals making $1 million a year while everyone else sees their respective earnings dwindle to between $5,000 and $10,000 per annum.[19] When asked whether people would accept this, Cowen responds that, hey, many people in Mexico are happy making much less. "They hardly qualify as well off but they do have access to cheap food and very cheap housing. Their lodging is satisfactory, if not spectacular, and of course the warmer weather helps." On Cowen's planet, http://www.leftoverswap.com would rule and Walter White, the antihero of *Breaking Bad,* might run an advice column inspired by the *Hunger Games.*

In related news, Uber recently "honored" schoolteachers who drive for UberX after hours to put food on the table.[20]

Tyler Cowen's vision plays well with the antiunion campaigns launched by Reagan and Thatcher in the 1980s, which touted individualism while reducing social spending, created deliberate shockwaves of austerity, and aimed to destroy the spirits of unionized flight traffic controllers and miners. Is digital labor primarily an instrument for the destruction of the middle class, a move to further politically disempower citizens?[21]

If companies like Work Market have it their way, they will become the middlemen when traditional employees at corporations like IBM are replaced by "independents." As the CEO of Work Market pointed out at the 2014 Share NYC conference, their five-year plan is to achieve a sweeping turn away from wage employment: shifting a market characterized by full employment—jobs for life—to a market that is "liberated" from employment, and dominated by freelancers, part-timers, and independent contractors. The big venture capital behind companies like Uber "influences" *(corrupts?)* the vision of worker rights that is being adopted by municipal law and American labor law more broadly, turning such labor companies into citadels of antiunionism.

Stand in Solidarity

Who stands in solidarity with the poorest, most exploited workers? Historically, capitalist owners were faced with a mass of workers, frequently represented by a union. But today workers are, in many cases, anonymous individuals facing off against anonymous employers. At present, unions cannot easily represent most workers through firm-by-firm collective bargaining: Workers often have contracts with more than one company at a time, and many of these companies are skeptical of unions.

What do we make of the steady decline of unions over the past sixty years? In 2012, union membership in the United States reached its lowest point in over one hundred years: Only 11.3 percent of public-sector employees and 6.6 percent of those in the private sector belonged to a union. The implosion of the Soviet bloc around 1989 was one reason for this decline; it removed the only living comparison to another social system and opened up global markets. The class of owners no longer needed to fear a mass exodus of organized workers who would descend on their suburban cottages.

Canadian sociologist Vincent Mosco, and of course the Wobblies before him, discussed the dream of one big union of unions, capable of converging various forms of solidarity worldwide.[22] Why couldn't American and Indian workers connect online and stand united as they face Amazon's CEO, Jeff Bezos?

American unions are too busy with their own problems to take on the plight of digital laborers. I am not aware of a single American union that has taken on the issue of digital labor explicitly. In contrast, the German Association of Unions proclaimed, in reference to "digital McJobs," that they "will not passively watch a modern form of slavery emerge, which drives competition to the bottom."[23] While such a reference to slavery is unwise, it is commendable that this association of unions has put the issue of exploitative digital work practices on the table. The largest German union, IG Metall, issued a press release suggesting that there has been a pervasive moral decline in the workplace due to digital labor.[24] In addition, they just published the edited volume *Crowdwork—zurück in die Zukunft.*[25] A society that wants to call itself a democracy should not tolerate workplace exploitation in any form.

Social Media for Worker Mobilization

These are not the days of *On the Waterfront,* a film showing how unions controlled the hiring on the New Jersey docks in the mid-twentieth century. Today's digital laborers cannot be reached in cafeterias during lunch break, or outside the gates of the factory. The character Terry Malloy, played by Marlon Brando, would instead be on LinkedIn. He could use apps like *LabourLeaks,* which, taking the spirit of *WikiLeaks* and Anonymous, calls on would-be whistle-blowers to publicize nuts-and-bolts accounts of underpaid and dangerous work. Or he could crowdfund and gamify worker organizations, handing out badges for talking to other workers instead of rewarding workplace efficiency. Terry Malloy could use Coworker.org, a platform for labor organizing that convenes around worker rights. Why not "napsterize" the Teamsters with peer-to-peer technology, anonymity gear like Tor, or LiquidFeedback, the free software tool for political opinion formation.

Where is the 4chan—the /b/ board—for viral labor memes? Workers at Foxconn in Shenzhen are using social networking platforms like Qzone and Renren to inform workers about union campaigns; but in doing so they are also making it easy for employers to identify them. Crabgrass would make their activist organizing safer, as it is especially designed for activist organizing.

"Tough day at work? Are you feeling overworked, underpaid, unsafe or disrespected by your boss? You are not alone—and you don't have to just put up with it."[26] The American Federation of Labor and Congress of Industrial Organizations (AFL-CIO) asked these questions and now offers its own toolkit, including the online platform http://www.fixmyjob .com/—also accessible through http://www.organizewith.us/. With the help of these tools, workers can clarify their grievances and discover possible routes for collective action.

The Italian media theorist Tiziana Terranova, in her contribution to The New School's Digital Labor conference in 2014, writes that "the old forms of blocking production are obsolete, if not impossible."[27] Instead of these forms, she reframes the traditional general strike as the "social strike"—a permanent experiment of invention that diffuses forms of striking practicable even by those who would, according to the traditional model, be incapable: the unemployed, the precarious, the domestic worker, the crowd worker, the migrant without official documents. The social strike aims to redeploy, reconnect, and reinvent all forms of strike: "The general strike of those who cannot strike, net strikers, strikers within the spaces of education, the gender strike."[28]

Electoral Politics and the Physical Backbone of Digital Labor Brokerages

To this kaleidoscope of strategies, add physical resistance and electoral politics, and then consider them in the context of digital labor. Take the great victory of 2012, when the Stop Online Piracy Act (SOPA) was defeated in the U.S. House of Representatives. SOPA's objective was to control and censor Internet users, to prevent them from violating the copyrights of third parties. Netizens cooperated with large businesses to force their point. On January 18, 2012, thousands of websites, including Google and Wikipedia, went dark or offline for twenty-four hours to make a clear point that such copyright enforcement would be a gross act of censorship. After receiving millions of e-mails, seeing street protests, and receiving countless phone calls and letters, elected officials in Washington realized that SOPA could become a voting issue. The SOPA example could inspire a new species of electoral politics, one led by associations and driven by the pursuit of worker demands. If enough people clamor for better working conditions—based on their own sense of dignity and the right to speak authoritatively of their own lived experience—this may sway some politicians.[29] In the case of SOPA, new media companies worked with activists to challenge the bill. But who will stand with the digital laborers? Perhaps the most promising response to this question would be the founding of worker cooperatives.

Inspiration could also come from the activist strategies of ACT UP, a coalition of AIDS activists formed in the 1980s when Reagan ignored the deaths of countless gay men who had contracted the immunodeficiency virus. Activists illegally entered the press offices of the firm that produced the murderously priced HIV drugs and faxed out press releases stating that the company would substantially lower the price of the drugs.

Digital labor brokers have headquarters, too—possibly in your city. Protests held right there—outside the offices of Bezos and Biewald— might be a fun and promising way to think about worker pushback. Amazon.com's headquarters is located at 1516 2nd Avenue in Seattle.

Cooperatives, Innovative Unions, Guilds, and Design Interventions

Innovative Unions

Freelancers Union, founded in the United States by Sara Horowitz in 2001, offers health insurance to each of its 250,000 members: temps, freelancers, part-timers, and other precarious workers who are not insured by their employer. Horowitz sees upsides of freelancing in the freedom from authoritarian workplaces, the autonomy to set one's own work schedule, and the freedom to make alliances with like-minded people. A setback for Freelancers Union has been the introduction of the Affordable Care Act, which led to a two-thousand-dollar average increase in annual premiums for members after the initial one-year waiver. Nevertheless, the union provides a client rating system, insurance plans, networking opportunities, and now also a primary care practice for freelancers in New York City.[30]

Many Turk workers are outspoken about their lack of interest in unions. Friends recommend to friends that they try out MTurk, and they recommend to each other better-paying tasks. But turkers also join worker-run forums like TurkerNation, Cloudmebaby, MTurkGrind, and the Reddit subreddit (/r/mturk) to chat, seek emotional support, and direct each other to "lucrative" tasks. On TurkerNation, workers express frustration with particular employers—but such disproval is clearly distinguished from any broad dismissal of Amazon, the company that sustains or contributes to their livelihood.

Design Interventions

Lilly Irani, a professor at UC San Diego, asks how to build a system that can support collective action online. How can you gather people, gain critical mass, and mobilize? Together with Niloufar Salehi, Michael Bernstein, Ali Al Khatib, and Eva Ogbe, she built a platform called Dynamo that allows workers to safely post and discuss ideas for actions. One such proposal was to start an e-mail campaign to Jeff Bezos with the intention of humanizing the workers on MTurk.

Irani is also the cocreator of a rating system, Turkopticon, which allows turkers to flag companies for bad behavior.[31] Irani's tool, a Firefox plug-in, is already used by roughly twenty-two thousand workers in any given month. Turkopticon, named a bit tongue-in-cheek after Jeremy Bentham's panopticon, is designed as a social support system for MTurk workers. It helps them to identify quasi-employers who don't pay, who severely underpay, or who don't respond to workers whose work has been rejected. If sufficient numbers of workers were to join the platform, employers might have to care about their reputation within the workforce. Design interventions like Turkopticon aim to bring fairness and social peace to platforms like MTurk.

Worker Cooperatives

Workers in Greece are now practicing a more structural approach. They are forming worker cooperatives, fab labs, and hacker spaces that sometimes collaborate with unions. In Athens, for example, there's the Δικτύωση Συνεργατικών Εγχειρημάτων Αθήνας (Networking Cooperative Ventures), which defines itself as "a collective of people who were either unemployed or worked in precarious employment. Faced with this reality, we decided to try another way of working—collective, based on respectful relationships, camaraderie and solidarity."[32] Worker cooperatives are owned, controlled, and self-managed by the workers themselves.

In *This Changes Everything*, Naomi Klein recounts her experience of living in Argentina for two years, making a documentary about workers who turned their old and abandoned factories into cooperatives following the country's economic crisis in 2001. Her documentary, titled *The Take*, tells the story of a group of workers who took over their shuttered auto parts plant and turned it into a thriving co-op. Workers took significant risks, but over a decade later, the factory is still going strong. In fact, the majority of worker-run cooperatives in Argentina—and there are hundreds of them—are still in production today.

New Guilds

In his book *The Precariat*, University of London professor and labor activist Guy Standing calls for new guilds to fight for more than just better working conditions. It is not sufficient to fight for higher wages; the very structures of production should be under scrutiny. Following the model of social movement unionism, guilds and associations could engage in wider political struggles for social justice and democracy. Digital labor associations, like TurkerNation, could coalesce with existing movements such as the National Domestic Workers Alliance and its equivalents in the fast-food industry. While there are, of course, vast differences, precariousness unites these groups. TurkerNation could build a worker-owned, apps-based labor platform. But beyond that, TurkerNation could fight for the recognition of invisible sites of work and obscure forms of employment more generally, and support campaigns for guaranteed basic income which would secure the future for crowd labor. In my opinion, such digital labor guilds could inform workers of their rights; challenge the status of workers as independent contractors through coordinated campaigns and class-action law suits; celebrate ethical crowdsourcing companies that pay a living wage to their workers; call for international codes of good practice and a restructuring of social protection for the contingent workforce (as advocated by Standing); document, as well as publicize, unfairness; lobby for the application and enforcement of federal labor law; and advocate for more time for thinking, dreaming, and imagining.

The Time Has Come for Platform Cooperatives

Earlier in this chapter, I referred to the possibility of an apps-based, worker-owned labor cooperative. Just for one moment, imagine that the algorithmic heart of any of the citadels of antiunionism, like Uber and TaskRabbit, could be cloned and brought back to life under a different ownership model, one featuring fair working conditions, a humane alternative to the free market model.

Just take Uber's app, with all its ride-ordering and geo-location capabilities. Why do its owners and shareholders have to be the main benefactors of such platform-based labor brokerage? Developers, in collaboration with local, worker-owner cooperatives, could design such a self-contained program for mobile phones. Despite its meteoric rise, three hundred million dollars in venture capital backing (and its evaluation bubble in the tens of billions), and massive international reach, Uber's long-term success is not inevitable. There is no magic sauce when it comes to developing such a piece of software; it's not rocket science. Of course, technology is only one part of the equation, and instead of letting techno-determinism run its course, I'd rather point to the long history of worker-owned cooperatives, as described or advocated by the likes of E. P. Thompson and Robert Owen.

Taxi drivers and technologists can coalesce to build an app that equals or outperforms its corporate equivalent. This movement has already started with a driver-owned ride rental service called La'Zooz and a cooperative version of eBay called Fairmondo. Worker-owned cooperatives can offer an alternative model of social organization capable of addressing financial instability. They will need to be collectively owned, democratically managed businesses that make it their mission to anchor jobs and offer health insurance, pension funds, and a degree of dignity. Mondragon, an often-cited example, is a fully incorporated federation of worker cooperatives that was founded in 1956 in the Basque region of Spain. It is worker-owned, not worker-managed, and functions within the larger competitive market.[33] At the end of 2013, it employed 74,061 people in the areas of finance, retail, and education. Mondragon cooperatives are united by a humanistic concept of business. The general manager of an average Mondragon cooperative makes no more than five times the minimum wage paid in his or her cooperative. (Compare that to Walmart's CEO, who is paid 1,034 times more than the median Walmart worker.)

But cooperatives also face many challenges on the level of competition from dominant players like Uber, and also in terms of public awareness, allocation of work, and wage levels.

Platform cooperativism equals a more humane workplace equals real benefits. It can invigorate genuine sharing and does not have to reject the market. Platform cooperativism can serve as a remedy to the corrosive effects of capitalism and it can be a reminder that work can dignify, rather than diminish, the human experience.

Platform cooperativism, innovative unions, the "social strike," electoral politics, ethical businesses, and design interventions are not panaceas, but they can start to weave some ethical threads into the fabric of twenty-first-century work. We need to take a long-term approach to the development of the digital labor movement, focusing not so much on the success of individual actions or tools, but rather on getting more people to join the ranks.

This fight, as the word suggests, will require a commitment to experimentation with new forms of solidarity. Stephen Duncombe's book *Dream* argues that progressives should be unashamed to wrest back modes of thinking that include ideal states of security and ease. Let's demand a world that we actually want to live in. Let's not take anything as unchangeable and let's be clear: Without conflict, outrage, and protest, structural change cannot emerge.

Notes

1. Trebor Scholz, "Digital Labor: Nov 14–16, NYC," Digitallabor.org, December 18, 2014, http://www.digitallabor.org/.

2. Karën Fort, Gilles Adda, and K. Bretonnel Cohen, "Amazon Mechanical Turk: Gold Mine or Coal Mine?" *Computational Linguistics* 37, no. 2 (2011): 413–20; Andreas Wilkens, "Digitale Arbeit: IG Metall sieht Sittenverfall," *Heise Online,* October 11, 2014, http://heise.de/-2390001/; Moshe Marvit, "How Crowdworkers Became the Ghosts in the Digital Machine," *The Nation,* February 4, 2014, http://www.thenation.com/article/178241/how-crowdworkers-became-ghosts-digital-machine/.

3. "NY Taxi Workers (@NYTWA)," Twitter, https://www.twitter.com/NYTWA/.

4. Amanda B. Johnson, "La'Zooz: The Decentralized Proof-of-Movement 'Uber' Unveiled," *The CoinTelegraph,* October 19, 2014, http://cointelegraph.com/news/112758/lazooz-the-decentralized-proof-of-movement-uber-unveiled/.

5. Fort et al., "Amazon Mechanical Turk"; Wilkens, "Digitale Arbeit"; Marvit, "How Crowdworkers Became the Ghosts in the Digital Machine."

6. Andrew Ross, *Creditocracy: And the Case for Debt Refusal* (New York: OR Books, 2014).

7. In 2012, 21.2 percent of New Yorkers were living below the poverty line, writes Sam Roberts in "Poverty Rate Is Up in New York City, and Income Gap Is Wide, Census Data Show," *The New York Times*, September 19, 2013, http://www .nytimes.com/2013/09/19/nyregion/poverty-rate-in-city-rises-to-21–2.html/. However, following a "living wage calculator" designed by MIT, the poverty line, defined by the federal government as $19,790 a year for a household of three, should in fact be between $46,000 and $67,000. This calculation would place far more families below the poverty line. http://livingwage.mit.edu.

8. Karën Fort et al., "Amazon Mechanical Turk"; Andreas Wilkens, "Digitale Arbeit"; Moshe Marvit, "How Crowdworkers Became the Ghosts in the Digital Machine."

9. Joel Ross et al., "Who Are the Crowdworkers?: Shifting Demographics in Amazon Mechanical Turk," in Proceedings of CHI 2010, Atlanta, Ga., ACM, April 10–16, 2010, 2863–72.

10. Fort et al., "Amazon Mechanical Turk."

11. Personal exchange with the author.

12. Massolution, "Enterprise Crowdsourcing Research Report," Crowdsourcing .org slide presentation, February 2012, http://www.crowdsourcing.org/document /enterprise-crowdsourcing-research-report-by-massolution-market-provider -and-worker-trends/13132/.

13. "Amazon.com Research and Development Expense (Quarterly) (AMZN)," *Ychart*, R&D expense statistics, http://www.ycharts.com/.

14. Greg Grandin, *Fordlandia: The Rise and Fall of Henry Ford's Forgotten Jungle City* (New York: Metropolitan Books, 2009), EPUB e-book, chapter 2.

15. Moshe Marvit, "How Crowdworkers Became the Ghosts in the Digital Machine."

16. William Thies, Aishwarya Ratan, and James Davis, "Paid Crowdsourcing as a Vehicle for Global Development," Microsoft Research, presented at the ACM CHI 2011 Workshop on Crowdsourcing and Human Computation, May 2011, http:// research.microsoft.com/pubs/147084/ThiesRatanDavid-Crowd-Final.pdf/.

17. "MobileWorks," Mobileworks.com, https://www.mobileworks.com/.

18. Fred Turner, *From Counterculture to Cyberculture: Stewart Brand, the Whole Earth Network, and the Rise of Digital Utopianism* (Chicago: University of Chicago Press, 2008), 9.

19. Tyler Cowen, *Average Is Over: Powering America beyond the Age of the Great Stagnation* (New York: Dutton Adult, 2013), EPUB e-book.

20. "As communities are heading back to school, we'd like to take a moment to celebrate the educators who are also our Uber partner drivers," a company representative, "Lindsey," stated in a blog post titled "Teachers: Driving Our Future," Uber Cincinnati Blog, September 4, 2014, http://blog.uber.com/UberTeacher/.

21. My coupling of inequality and political disempowerment is inspired by a conversation with Karen Gregory. For most people, inequality is not solely about being poor; it also entails not having a voice in political life.

22. Fort et al., "Amazon Mechanical Turk"; Wilkens, "Digitale Arbeit"; Marvit, "How Crowdworkers Became the Ghosts in the Digital Machine."

23. Ibid.

24. Andreas Wilkens, "Digitale Arbeit."

25. When translated, the title is "Crowd Work—Back into the Future." Christiane Benner, *Crowdwork—zurück in die Zukunft* (Frankfurt am Main: Bund-Verlag GmbH, 2014).

26. "Fix My Job," *Working America*, accessed December 19, 2014, http://www.fixmyjob.com/.

27. Terranova, "Social Unionism."

28. Fort et al., "Amazon Mechanical Turk"; Wilkens, "Digitale Arbeit"; Marvit, "How Crowdworkers Became the Ghosts in the Digital Machine."

29. Another relevant reference is the Party of the Poor, a short-lived left-wing political movement and militant group that operated in Mexico between 1967 and 1974.

30. Thornton McEnery, "Freelancers Union Retools for Obamacare," *Crain's New York Business,* http://www.crainsnewyork.com/article/20140224/HEALTH_CARE/302239986/freelancers-union-retools-for-obamacare/.

31. "Turkopticon," ucsd.edu, http://turkopticon.ucsd.edu/; "DYNAMO," Wearedynamo.org, http://www.wearedynamo.org/.

32. Δικτύωση Συνεργατικών Εγχειρημάτων Αθήνας. Kolektives.org, http://www.kolektives.org/.

33. George Cheney, *Values at Work: Employee Participation Meets Market Pressure at Mondragon* (Ithaca, N.Y.: ILR Books/Cornell University Press, 1999).

· II ·

Openness

CHAPTER 5

Paradoxes of Participation

Christina Dunbar-Hester

O NE AFTERNOON during a weekend workshop where volunteers were building a new low-power FM radio station, a middle-aged electrician approached me and apologized for making me cry. This was puzzling to me, because I had not interacted with him at any point. I looked at him quizzically, and he quickly realized his error: He had mistaken me for another young white woman with short dark hair. Naturally I wondered what was going on. We figured out he had thought I was a volunteer named Louisa, and he asked me to tell her that he was looking for her, if I saw her. A few hours later, I bumped into Louisa, and alerted her that she was being sought.[1] She declined to explain the situation in the moment, but later in an interview she briefly described what had happened:

> I tried to get involved in some carpentry [to build the radio studio]. And I didn't understand what [the electrician] was saying, and I just . . . walked away. . . . But I [had] really wanted to be a part of the carpentry and I wanted to learn and I wanted to get involved.
> [Specifically,] he was talking about some kind of nail, and measuring from this point to that point, and I was kind of like, "which point again?" and he got snappish and was just like "just let me do it!" And once you start with the "just let me do its," you don't feel welcome and you don't want to be involved.[2]

I was present at this gathering, held in rural Tennessee in the spring of 2005, in my capacity as an ethnographer. I was studying the politics of technology in media activism by "deeply hanging out"[3] with a group of Philadelphia-based activists who, among other activities, traveled the country building new micropower radio stations at events like this workshop. As it happened, while immersing myself in their activities I was inadvertently drawn into the conflict related above, which hinged specifically on matters of pedagogy, novice versus expert status, gender, and technical familiarity. Though my point of entry into this situation was unusual—I was not even involved in the misunderstanding—these sorts of issues were not uncommon amid an ethos where both technical skill and novice participation were prized.

The reality of expertise ran afoul of the activists' exaltation of technical participation. A major plank of the radio activists' work was the promotion of technical participation to novices through various activities such as radio station–building workshops, tinkering meet-ups, and other types of DIY (do-it-yourself)[4] work with technology.[5] They routinely presented the work of soldering a transmitter, building an electronics console (as in the carpentry example above), or tuning an antenna to be accessible to all. They invited novices to participate in these activities, to "put their hands on the technology," and held that such experiences in technical participation were liberating. Specifically, the radio activists sought to offer "participation" as an experience to everyday people. They presented technical engagement as a strategy not only for leveling expertise but for increasing political participation as well. They believed that technical work could impart a heightened sense of agency to participants. They recognized that tinkering is as much a form of cultural production as a technical one.[6]

This chapter examines how activist ideals manifest in the realm of practice. To do so, it follows the work of a group of media activists whose work foregrounded engagement with communication technologies. As demonstrated by the anecdote above and a short ethnographic vignette below, the activists, working in a self-consciously collaborative mode, promoted hands-on work with radio hardware as a means of enacting participatory politics. This practice was understood to be in the service of a broader goal of facilitating technical and political engagement through a "demystification" of technology.

The promotion of technical participation as a route to wider empowerment reveals two paradoxes of participation, both of which are foreshadowed above. Louisa's informal and ad hoc attempt to "plug in" to the carpentry work in the radio studio was a result of the radio activists' deliberate choice to leave much of the work of building the radio station relatively unstructured. Their vision for "participation" included the experience of self-guided discovery and learning, and the formation of an affective connection to activism through the practice of making one's own way. Ironically, though, such a putatively self-directed exploration of new skills can back-fire. The radio activists' participatory ideal left Louisa, a novice, dependent on the electrician building the console to teach her about carpentry and electronics. It also made the electrician responsible for helping Louisa move from her absolute beginner status toward a burgeoning sense of technical engagement. But the electrician was not only trying to engage Louisa technically and pedagogically according to the activists' vision; he was also trying to build a working console in a compressed amount of time. This circumstance raised the stakes of his and Louisa's encounter, putting their goals at odds with each other, which in turn undermined Louisa's ability to explore and learn. Thus, a "participatory" mode that made peers responsible for producing the console together was itself responsible for frustrating and alienating a novice to the point of tears.

A second paradox is revealed in how there were patterned gaps in the radio activist organization and volunteer base that undercut the activists' commitment to egalitarian participation. Louisa and the electrician came together at the workshop bearing the full weight of their social identities, which preceded and included relationships with technology. Men were more likely than women to know how to use the tools the activists touted, to build electronics, to be excited by tinkering, and to have the know-how to teach neophytes. This troubled the activists, who fervently hoped to provide a participatory experience that was universally attainable; the last thing they wished to do was to reproduce a hierarchy of technical partici-pation based on gender roles. But technical "participation" was vexed by the gendered legacy of the activities that the activists prized. Women, by contrast, were more likely to exhibit comfort with cooking, cleaning, and managing logistical work. In order for them to move toward technical par-ticipation, they had not only to take up new skills and tools but also to leave behind familiar activities—a daunting proposition when *all* of these activities, from soldering to cooking, needed to be accomplished at the activists' worksite. The activists' greater attention to the technical side of work left them unprepared to address gender parity across the multiple domains of work that their practice encompassed.

Background

The activists whose work is presented in this chapter came together in the mid-1990s as a pirate radio collective in Philadelphia, Pennsylvania. After they were raided and shut down by the Federal Communications Commission in 1998, they turned away from broadcasting and toward policy advocacy and building radio stations. Expanding access to low-power FM radio (LPFM) was their main concern, but in the early 2000s they also considered whether and how to expand their mission to "free the airwaves" to include not only radio but also Internet-based technologies, especially community Wi-Fi.[7] They espoused radical leftist politics and considered their work as occurring against the backdrop of broader social movements for media democracy and social change.[8] The data in this chapter are drawn from a much larger ethnographic project, including participant observation and around thirty semistructured interviews, conducted between 2003 and 2007.[9]

The group's activities encompassed both advocacy to change policy (not discussed in this chapter) and assisting citizens and community groups with hands-on work with technology, including building new radio stations.[10] Technical engagement holds a special symbolic value across a diverse repertoire of activist practices. The activists convened weekly tinkering groups to build or repair electronics hardware, held other tinkering workshops like the transmitter workshop described below, and hosted radio station "barn raising," events where participants put a new radio station on the air over the course of a weekend. Barn raisings were highly symbolic events where the radio activists reinforced their twin missions of community radio and community organizing. The barn raising concept was a self-conscious reference to the Amish practice of people joining together to accomplish a project that an individual or small group alone would struggle to achieve, thereby emphasizing interdependence and cooperation.

Radio activism in this era must be understood as issuing from distinct yet interwoven social, cultural, technical, and political strands. These include embedded practices of community media production and pirate radio; "Indymedia" and the transnational anticorporate globalization movement;[11] the emergence of "new media" including the Internet; and a regulatory environment favoring national broadcasting networks and corporate media consolidation that was opposed by a growing movement for media democracy.[12] Other antecedents to radio activism include ham and citizens band radio, the appropriate technology movement of the 1960s and 1970s, and earlier broadcast reform movements.[13]

Technical Participation in/as Practice

The activists convened another technical workshop that spring on their home turf in Philadelphia. It was organized like a "mini-barnraising" and was oriented around diagnosing and repairing two large radio transmitters from the 1970s that had been donated to the activists by a college radio station in upstate New York. Decommissioned when the station upgraded its equipment, the transmitters were in poor condition, and they were also filthy. They were essentially trash, albeit specialized electronic trash that was of keen interest to activists who valued reuse, keeping old technology alive, and the learning and teaching potential of even nonfunctional electronics hardware. Each of the transmitters was the size of a refrigerator (see Figure 5.1). They were not in working order and had been out of use for decades. They were heavy to move, difficult to see into, dirty inside and out, and missing various components. If operational, the signal of one of the transmitters would have been 10,000 watts; the capacity of the other was 1,000 watts.[14]

Figure 5.1. A staff activist and two volunteers unloading a transmitter from a truck. Photo courtesy of Prometheus Radio Project.

Like a barn raising, the workshop featured explicit teaching tracks running alongside constant work on the transmitters, and people moved fluidly to drop in and out of formal and informal activities. As in a barn raising, the whole group broke for meals together. There were fifteen to twenty-five participants for most of the weekend. These included four to five paid staff members of the activist organization; their interns (at any given moment, the group had a rotating cast of two or three and their internships would last a semester, summer, or academic year); novice volunteers (from Philadelphia and New York City, most of whom had paid a nominal fee to participate in the workshop and learn about hardware); four to five highly skilled engineers who the radio activists had enlisted to help troubleshoot and teach (some local, some from as far away as Washington state and Illinois); and myself as a participant observer. The engineers and volunteers held different sorts of day jobs—some in community media, some as engineers, and some in unrelated fields. All of the engineers helped build community radio stations on a regular basis, though in most cases on an unpaid, voluntary basis; most had formal engineering training of some sort, and the deepest technical expertise in the group resided with them.

The engineers focused on cleaning, assessing, and diagnosing the hardware, while staff activists ran lectures and tutorials for the novices. Volunteers and staff drifted from the formal workshops into the truck in order to clean, ask questions, or simply watch what was going on, and the engineers would sometimes work on removable parts outside at tables in full view, and make attempts at explaining what they were doing. Since the workshop was in the public outdoors and the weather was pleasant, the activists set up a table with brochures, chatted with passersby about the workshop and the organization in general, and solicited donations. The truck was festooned with a sign that read, "What are these crazy people doing inside that ginormous truck? Come in and find out!" This improvised publicity represented the activists' symbolic goal of expanding participation; they would have eagerly welcomed neophytes off the street.

On Saturday, staff activist Jasper led a teaching track about the technical properties of radio, providing an overview of the physical properties of radio, electromagnetism, and hardware. The track was attended by the novices in the group. Simultaneously, people worked to clean the transmitters with rags and toothbrushes, and to perform diagnostic work dominated by the most experienced engineers. Early on, the engineers deter-mined that the higher-power transmitter was in better condition than the 1,000-watt one, so effort was focused on the 10,000-watt machine. Jasper himself had a deeper engineering background than the volunteer work-shop attendees, but was largely self-taught, and was less expert than the engineers or Brian, the only staff activist with formal engineering training. Jasper's lecture included an introduction to the parts of a radio station, antennas and standing wave ratio, the electronic components found in a transmitter, and power, moving between political and technical registers and even punning to connect them. He also displayed the activists' ideal-ized model of expertise, stressing that he, too, had recently been a novice and taking pains to promote egalitarianism in technical practice: "One of the good things about me teaching you is that I don't really know that much about radio. I'm not that far ahead of you, as opposed to people who know way more and are basically incomprehensible."[15] He explained resis-tors as follows:

> This is a good word for radicals who are against the state. [laughter]
> It's measured in ohms. Think about water. The bigger the tube, the
> less resistance it encounters as it goes through the tube. Resistance
> is not in itself a bad thing, sometimes in a circuit there are advan-
> tages to not letting all the power flow. A light bulb is a resistor, it
> makes electricity flow slowly and heat up the filament and turn it
> into light.[16]

In tying political radicalism to ohms, his statement conjoins a political stance to technical affinity, in keeping with the activists' wider political project.

In an interview, another activist named Brian reflected on agency and expertise:

> You can do any tech project . . . you can do this stuff and you can
> self-educate. . . . Culturally we have a very expert-oriented society . . .
> you have all these people who are "experts," and just because
> they're talking at you about these different things, doesn't necessar-
> ily mean they're right. . . . The big part . . . about not having the
> engineers do it, it is a demystification, and making people feel like,
> oh, experts just happen to know this, they've just done this a

bunch, giving people the feeling, oh, if I just did this enough, I could do this just as well as this guy, as well as this engineer.[17]

The activists promoted their vision, which included a demystification of technology, participation by novices, and the leveling of technical expertise through pedagogical activities. But the ideals promoted by Jasper and Brian ("you [too] can do any tech project") represented fantasy. In fact, many technical projects remain inaccessible without years of training. Furthermore, informal education offered by volunteer or activist associations is often limited due to the way in which they run on donated time, and moreover conduct training outside of formal work or educational activities. That being said, radio is a technology that works relatively well with this fantasy: It is more common and less abstruse than many other technologies.

The equipment was filthy. Nearly everyone took a turn over the course of two days scrubbing inside the cabinets that held the components. Delicate or particularly dirty pieces were removed for special cleaning. A silver-plated vacuum tube had to be dusted and polished. It came out nicely. The most important diagnostic task was to see if the exciter worked. The exciter is the part of the transmitter that produces RF (radio frequency), and it can be assessed with tools and instruments that run on ordinary house current, since it only needs to put out around 300 watts. Other components could amplify this to 10,000 watts—though not at this workshop, as a generator to power up the big transmitters had been deemed a superfluous hassle and expense. Most participants were relieved by this. One of the engineers, Jim, warned everyone: "These big transmitters are dangerous. They must be used with respect. No one should ever repair, maintain, or even turn them on alone. They are deadly and you need another person to push you away if you start to fry!"[18] The task of diagnosing the exciter was largely dominated by the engineers, and the novices did not participate other than by hovering around and asking the engineers a few questions. At the end of the first day, Jasper and Brian asked the engineers to describe their labors to the group.

The second day of the workshop was less structured than the first. There were no lectures or formalized activities on Sunday. Cleaning, testing, and tinkering with equipment continued. Novice participants found themselves restricted to either cleaning components or helping with meals (see Figure 5.2). None too pleased with this division of labor, they took

Figure 5.2. Open transmitter cabinet with parts removed for cleaning. Cleaning was a way for novice young women to contribute "productively" to the effort to rehabilitate the transmitters. But it did not aid them in advancing their own technical expertise; in fact, it reinforced a gendered division of labor between those with the most expertise and novices. Photo courtesy of Prometheus Radio Project.

breaks in which they sat around chatting with one another and ruing the fact that they did not know how to "plug into" the technical work and did not feel especially welcome to do so.[19]

Thus the activists' desire to promote wider participation in technical practice was difficult to implement. Jasper, who had tried much harder than the engineers to make himself seem less expert and more accessible (as demonstrated, for example, by his lecture), was critical of himself and the activist organization for not trying harder to implement the stated barn raising ideal of "no one is allowed to do anything s/he already knows how to do" over the weekend. The transmitter workshop was a special "one-off" event in some regards; less planning had gone into it, by far, than into actual barn raisings. However, it was not unique in that it combined some formal structure with a strong self-organizing element. And yet the experience of the weekend amply demonstrated that without aggressive measures to combat the hoarding of expert knowledge (deliberate or not), the activists' vision for ecumenical skill-sharing could not be realized.

Jasper and Brian experienced a special tension, as they had much more technical knowledge than novices and interns but were also less expert than the visiting engineers. They were torn between trying to learn more themselves and extending their own understanding of technical problems on the one hand, and making sure that the engineers included the novices on the other hand. They desired to do both, but these goals were at odds with each other. They both repeatedly stopped the engineers to ask them to explain what they were doing while they were doing it, as well as insisting on accessible and public work reports at the end of the day. Brian in particular had a gentle yet persistent manner and would not permit the engineers to brush off his inquiries or his insistence that they explain their activities to the group and answer questions.

Nonetheless, for a group that contained expert members and that needed to accomplish many tasks, giving its novice members a full and comprehensive understanding would have impeded its engineer members' abilities to learn as much as they could about what was wrong with the transmitter, and the engineers were not terribly interested in slowing their work down to explain it, let alone give over the equipment and diagnostic tools to novices (see Figure 5.3). And the novices pausing their cleaning activities to learn more from the engineers would have prevented the massive cleaning undertaking from getting as far as it did; novices mostly stuck to what they knew they could do, and did not feel inclined to cease being "productive" themselves, nor to interrupt the engineers. Hence technical nonexperts primarily cleaned and provided meals, while technical experts primarily performed tasks that required electronics expertise.

At the end of the weekend, the transmitters were not repaired and they needed to go back into storage (the exciter worked, but on its own it was of little practical value). Plans were made to bring the engineers back to have another go at the project, probably with a generator. Notably, this was not planned as another pedagogical workshop, and would probably involve a more expert group focused on getting the transmitters running rather than skill-sharing. This decision represented an acknowledgment of the uphill battle of supporting egalitarian technical practice, and indeed the failure of the group to fully implement certain ideals in this case.

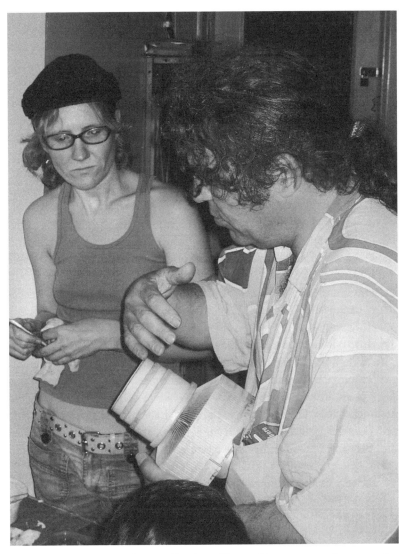

Figure 5.3. Engineers removed potentially salvageable parts like an exciter and vacuum tube. Here, an engineer shows the vacuum tube to a volunteer. While engineers advanced the repair mission by sequestering working parts for diagnostics, novices were relegated to tasks that did not require a great deal of technical know-how. Photo courtesy of Prometheus Radio Project.

To reprise, the activists' stated participatory ideal was that "no one is allowed to do what they already know how to do" at a barn raising. Activists attempted to salt all technical undertakings with pedagogy: Expert engineers and activists were supposed to guide novice volunteers through the assembly of the new radio station, handing tools off to other people to learn new skills. This was seen as an exercise in community empowerment, and technical practices were explicitly linked to political engagement. A staff activist reflected on this, invoking DIY: "[A] big part of the barn raisings [is that] it is a demystification, and making people feel like . . . oh, if I just did this enough, I could do this just as well . . . as this engineer."[20] In practice, though, this ideal remained out of reach; there were formidable barriers to leveling technical expertise.

Moreover, if we turn back to Louisa's experience and that of the novices at the transmitter workshop, we will note that this exaltation of "participation for all" glosses over historical reality. Radio tinkering was established as a masculine pursuit in the early twentieth century, and remained understood as an activity pursued by elite boys and men for many decades.[21] As Susan Douglas writes, "The course of radio's early development was . . . influenced by the professional aspirations and leisure activities of a subculture of middle-class men and boys."[22] Far from being "for all" in its early instantiations, tinkering as a hobby often bled into formal employment in technical fields. Hobbyist technical practice was a focal point around which the masculine, white-collar, and elite status of technical occupations became entrenched. Later, the electronics tinkering practices that had crystallized around radio shifted to include computers and programming, but the association of these activities with white and white-collar masculinity was strongly rooted.

The radio activists hoped to challenge these associations by presenting technical affinity as universally attainable, but this approach overlooked the gaps in existing skill and affinity among their volunteer base and the members of the public to whom barn raisings and related workshops were addressed. Some people came to radio activism with already developed skills and enthusiasm for technical work, while others found these practices foreign. Technical pursuits were fun for some people (especially men),[23] but potentially intimidating or unappealing for others. This was true even though the activists self-consciously tried to distance themselves and their pedagogical practices from the competitive and exclusionary aspects of some engineering and electronics cultures.[24]

Nonetheless, the historical legacy of electronics employment and hobbyist cultures loomed large. It meant that leadership and teaching work was most often performed by expert men. Women were more often novices. Members of each gender thus experienced unique pressures. As discussed above, the engineers at the transmitter workshop and the electrician at the barn raising were torn between making measurable progress on their technical tasks and slowing down to train novices and allowing them to "put their hands on the tech." In addition, people who found tinkering and problem solving affectively pleasurable were not necessarily as gratified by ceding control over tools and equipment to novices. The staff activists were more accustomed to working within these parameters, and more committed to "opening up" technology in practice, but the engineers in particular had to be reined in at points.

The burden for women and novices was even greater. Their attention was divided between being useful to the overall effort and taking up the practices suggested by the activists' vision. The activists promoted the ideal that "no one does what s/he knows how to do already," giving novices a point of entry into technical work. But this also caused novices to feel conflict over whether their efforts were productive. When people stuck to what they *did* know how to do, more progress occurred. For example, Louisa explained to me that she had sought out the carpentry work only after being relieved from hours on her hands and knees keeping people from tracking mud (a constant feature of the barn raising, as much of it was held outdoors and it had rained for two straight days) into an indoor, carpeted space. She was interested in carpentry specifically because it was unfamiliar to her and because she was feeling burnt out by the logistical work to which she had been assigned. But even though she was exhausted from it, she felt that she could only abandon the mud-policing when someone else could pick up the task. From there, she threw herself directly into a challenging and novel environment, joining the electrician in the studio-in-progress. But as she acknowledged, by this point she was already fairly exhausted from having contributed her efforts to the organizational work that she had the ability to do.

Similarly, in the transmitter workshop, novices felt they could "productively" help by cleaning the transmitters or preparing meals. But they did not feel that they could effectively shift toward the more arcane tasks related to repairing the transmitters. Their expertise was too scant and they were hesitant to "interrupt" the engineers in order to enact the full pedagogical vision of the activists. They presumed, rightly, that their insistence on immersing themselves in the technical work would actually hinder its progress, if progress was defined as moving toward functioning hardware. While the activists' mandate to "participate" technically applied to everyone, the *burden* of participation fell disproportionately on women and technical novices. They experienced discomfort along multiple lines, including the feeling that to advance their own learning would not only set back the "technical" progress but also prevent other, less specialized work from occurring. Surfaces would get muddy, meals would not get cooked, the transmitter components would remain caked in filth, and so on.

It should also be noted that "no one doing what they know how to do" could be interpreted to mean that the engineers needed to pitch in with the cooking. But in practice, this did not occur, for two reasons. First, the valorization of technical work as the most prized enactment of the participatory ideal meant that mundane tasks (characterized by Louisa in the interview as "women's work") were rendered less visible; they were not constructed as liberating or politically significant in the way that technical work was. Second, the engineers alone possessed the expertise to get the hardware working: Without their efforts in this domain, a new radio station would not get built and the old transmitters would remain broken forever.

Conclusions: Participatory Problems and Potentials

The transmitter workshop contained elements of contradiction in its organization: It was self-organizing, though parts of it were also formally structured. Where did these opposing impulses come from? And what consequences were there for both the material and affective outcomes of the workshop?

The radio activists inherited elements of their practice from the appropriate technology movement of the 1960s and 1970s. According to historian Carroll Pursell, appropriate technology had origins in "the convergence of a broad countercultural movement, a reassertion of doubts about the role of technology in American life, and the burgeoning environmental movement."[25] These connections were strikingly apparent in the older engineers at the transmitter workshop. They were interested in carpentry and alternative energy as well as electronics and communication technologies. Ranging in age from their forties through their sixties, some were old enough to have perhaps cut their teeth in the original appropriate technology movement. As Pursell notes, a central claim of the appropriate technology movement was that these technologies "worked in gentle partnership with nature and fostered intimate personal relationships."[26] This idea resonates with the radio activists' notions about the community-level suitability of radio and its ability to foster transformative connections between neighbors.

This heritage played out in complex ways and had multiple implications for the transmitter workshop. Given their emphasis on personal and societal transformation, the activists did not wish to deny participants the experience of self-guided discovery, self-expression, or the formation of affective connection by controlling the workshop too tightly. On the other hand, if participants felt too impotent (or that what they produced was too inchoate), activists risked participants feeling as though their efforts had been wasted. Perhaps ironically, the transformative effects that were presumed to flow from technical engagement (imagined by both appropriate technology and participatory culture; see below) were most elusive when neophytes were denied a structured experience in engaging with the technology.

Especially on the first day, the radio activists cultivated structure. As previously described, they offered a formal teaching track in which novices were given a basic introduction to how a radio station works, with an emphasis on the technical aspects of broadcasting. In addition, Brian and Jasper took pains to interrupt the engineers for formal reports about their diagnostic activities. They insisted on recaps that included time for novices to ask questions (including very remedial ones). This move toward structure kept the novices occupied and included, providing them with tasks to perform and roles to play. It also offered the novices opportunities to speak and participate without fear of being judged as ignorant or a hindrance to the diagnostic and repair mission that was the ostensible goal of the workshop.

At the same time, structure was inimical to other goals of the workshop. On the most basic level, the radio activists sought to provide participants with transformational experiences.[27] The imposition of structure could potentially render the activists not accountable to participants' interests or values (especially to their exploration or creative expression). At the same time, a lack of structure ran the risk of producing disabling chaos or preventing the experience from having enough coherence to enable purposive engagement and acculturation. Novices were most frustrated when they were without prompts; "doing it themselves" when they lacked expertise was not ideal for them. And yet the radio activists were loath to impose too much coercive control over the event. This would have seemed to run against many of their organizational values and strong collectivist ethos.

Peer production (or participatory culture) is also highly relevant here as a related mode of cultural mediation. The radio activists' workshop possessed features that made it distinct from digitally networked peer production.[28] Namely, it was not digitally networked, distributed practice (though some elements of media activism are); rather, it occurred face-to-face. But its contours otherwise strongly resembled some of the features scholars of peer production have named as most significant. In particular, the workshop represented nonmarket and nonproprietary collaborative practice.

Two prominent claims about peer production are that it is especially egalitarian and especially gratifying for participants.[29] However, the radio activism example shows that some of what proponents have tended to assume about peer production is less evident in practice, along both of those lines. Another shortcoming of what Kreiss et al. term the "peer production consensus" is that it masks the fact that the dynamics of peer production may vary widely by site; open source software projects, for example, have traditionally been less committed to the participation of technical beginners, resulting in very different dynamics than those experienced by the radio activists.[30] Yet this workshop manifested aspects of participatory culture, including its self-organizing bent, mentoring/pedagogical dynamics, and cultivation of affective ties between members and between members and projects.[31] The radio activists' case is particularly illuminating for considering the interplay between technical expertise and an activist politics of technology devoted to "participation."[32]

Collective collaboration and the valorization of "participation" did not solve the "problem" of hierarchical organization that is often assumed to be a feature of bureaucracies but not of peer production networks.[33] Nor did this mode of practice confer an automatic sense of gratification on all participants. The novices' experiences show that there are good reasons to be wary of romanticized notions of voluntarism and participation, perhaps especially in the realm of technology. Novices needed guidance but could not easily shed their novice status.

Expertise was a significant issue in the interplay between structure and emergence. In an interview, Brian was critical of the culture of exclusion traditionally prevalent in engineering. He summarized his occasional attempts to "manage" engineers working on the activists' technical projects: "While I don't explicitly say, 'Stop being a patronizing asshole,' I have tried to communicate that."[34] Yet the problem went beyond merely keeping engineers from turning off novices by "being patronizing assholes." Indeed, the staff activists, not to mention the novices, needed the engineers if they were to make headway with arcane technical problems such as those they faced with the broken transmitters. Though activists valued self-organization and nonhierarchical participation, differentially distributed technical expertise threatened to exclude novices and erode the potential for "collaboration" that the activists embraced.

Last but not least, the complexity of gender and social identity as they intersected with the exaltation of technical participation presented distinct challenges. The legacy of electronics tinkering as a site where masculinity was constructed and reinforced meant that technical skill and affinity was unequally distributed within the activists' volunteer base and staff. The greatest concentration of expertise resided with a few expert men. When women and technical novices attempted to plug in to technical participation and put their hands on the radio technology per the radio activists' prescription, they not only confronted the legacy of their own exclusion but felt torn about neglecting the areas of work in which they were sure they could contribute. And the divide between novice participants and those who were deeply familiar with electronics (including some with formal engineering training) was not easily overcome by a simple prescription to include novices or bar participants from doing anything they already knew how to do; the technical and affective training this proposition required could not be imparted over a weekend. In spite of the activists' fervor for "demystifying" technology, technical participation as a route to more egalitarian social relations was less effective than hoped.

Rather than simply deeming these efforts a failure, however, we might take these episodes in participation as occasions to reflect on the contradictions between participatory politics and technical cultures predicated on elite forms of practice. The impulse to provide opportunities to attain heightened expertise in domains often closed to some people is admirable; the radio activists' attention to this disparity is commendable. At the same time, placing novices on a path of navigating their own way among experts here inadvertently reinforced hierarchies. It undercut the radio activists' mission to expand technical participation and placed a unique burden on those who already had the least familiarity with and affinity for technical work.

The ethos of "participation" configures acts of production as fun, transformative, informal, and ad hoc. But in fact, such notions belie the expert nature of technical knowledge. In practice, to raise technical novices up out of novice status would require more intensive resources and structure than the radio activists provided. The issues here are twofold: First, groups relying on voluntarism to accomplish labor, especially arcane labor, are plagued by the structural constraints of people donating spare time. The active mentorship required to bring novices to a greater familiarity and affinity for electronics work, for example, is difficult to accomplish when both experts and novices are dropping in and out of technical activities. Second, the ideology of participation serves to downplay the need for structure and resources.[35]

Thus, to leave the specific contours of "participation" incipient and determined by rushed volunteers on the fly was to miss opportunities to engage novices technically. But arguably, the exaltation of *technical* participation was itself part of the problem. Glossing over the more mundane practices also required to run a successful activist workshop (i.e., cleaning) gave short shrift to the full range of expertise and competence that volunteers—technically expert and otherwise—brought to bear on activist pursuits.

Notes

The author wishes to thank Biella Coleman and Jonathan Sterne for editorial comments on this paper.

1. The names used throughout this chapter are pseudonyms.

2. Interview July 25, 2006.

3. Clifford Geertz, "Deep Hanging Out," *New York Review of Books,* October 22, 1998.

4. DIY has at least two points of origin: first, as a project of masculine home improvement that carved out a masculine domestic domain in an otherwise feminized one; and second, within punk and hardcore music subcultures that called for resistance to the "appropriative and controlling" impulses of the commercial music industry. See Steven Gelber, "Do-It-Yourself: Constructing, Repairing, and Maintaining Domestic Masculinity," *American Quarterly* 49, no. 1 (1997): 66–112;

and Steve Waksman, "California Noise: Tinkering with Hardcore and Heavy Metal in Southern California," *Social Studies of Science* 34 (2004): 675–702. Mimi Thi Nguyen discusses the politics of race and gender in punk rock in "Tales of an Asiatic Geek Girl: Slant from Paper to Pixels," in *Appropriating Technology: Vernacular Science and Social Power*, ed. Ron Eglash et al. (Minneapolis: University of Minnesota Press, 2004), 177–90.

5. It should be noted that "technology" is largely an actors' category: The activists understood "technical" to refer to audio, computer, and radio transmission hardware, including electronics and carpentry tools, and software related to the production of community media. (Despite some differences, electronics practice should here be understood as being on a continuum with carpentry insofar as both historically occurred in settings like the "ham shack," a domestic masculine workspace carved out by men in contradistinction to home spaces shared with or controlled by women. See Kristen Haring, *Ham Radio's Technical Culture* [Cambridge, Mass.: MIT Press, 2006].) I do not mean to imply that other forms of interaction with artifacts or techniques are not "technical," but for the sake of the argument presented in this chapter I restrict the use of the term "technical" to the tools and artifacts related to broadcasting and media production.

6. See Carolyn Marvin, *When Old Technologies Were New* (New York: Oxford University Press, 1988), 7.

7. Christina Dunbar-Hester, "'Free the Spectrum!' Activist Encounters with Old and New Media Technology," *New Media & Society* 11, nos. 1–2 (2009): 221–40.

8. Notably, groups across the political spectrum have weighed in on media issues, especially to oppose media consolidation; it would be misleading to represent all groups engaged in media activism as having leftist politics. It would also be inaccurate to represent electronics tinkering as necessarily linked to politics or to leftist politics.

9. See Christina Dunbar-Hester, *Low Power to the People* (Cambridge, Mass.: MIT Press, 2014).

10. Of course, "citizen" is rightly a contentious concept for some. In my use of the term, I wish to signify activity around civic or communal participation, not to marginalize those without full legal status as citizens. (This is important as many media activists have a wider social justice orientation, including immigration rights. Several of the low-power radio stations that this activist group built were with migrant farmworkers' groups, as well.) Though I do not have space to interrogate "citizenship" here, using it to stand in for a mode of engagement open to "everyone" presents obvious problems.

11. Jeffrey Juris, *Networking Futures: The Movements against Corporate Globalization* (Durham, N.C.: Duke University Press, 2008); Todd Wolfson, *Digital Rebellion: The Birth of the Cyber Left* (Urbana: University of Illinois Press, 2014).

12. Robert McChesney, "Media Policy Goes to Main Street: The Uprising of 2003," *Communication Review* 7 (2004): 223–58.

13. Regarding ham and citizen-band radio, see Kristen Haring, *Ham Radio's Technical Culture*, and Art Blake, "Audible Citizenship and Audiomobility: Race, Technology, and CB Radio," *American Quarterly* 63 (2011): 531–53; regarding appropriate technology, see Carroll Pursell, "The Rise and Fall of the Appropriate Technology Movement in the United States, 1965–1985," *Technology & Culture* 34 (1993): 629–37; and regarding earlier broadcast reforms see Robert Horwitz, "Broadcast Reform Revisited: Reverend Everett C. Parker and the 'Standing' Case (*Office of Communication of the United Church of Christ v. Federal Communications Commission*)," *The Communication Review* 2 (1997): 311–48.

14. Machines running on so much power stood in marked contrast to LPFM transmitters, which by law cannot exceed 100 watts (about the same amount of power as an incandescent light bulb), with which the radio activists commonly worked. The big transmitters were unfamiliar to the core activists due to their power and scale, as well as not being solid-state.

15. Fieldnotes, May 28, 2005.

16. Ibid.

17. Interview, July 5, 2006.

18. Fieldnotes, May 28, 2005.

19. Fieldnotes, May 29, 2005.

20. Interview, July 2006.

21. Susan Douglas, *Inventing American Broadcasting* (Baltimore, Md.: Johns Hopkins University Press, 1987), chap. 6; Kristen Haring, *Ham Radio's Technical Culture*.

22. Douglas, *Inventing American Broadcasting*, xxii.

23. Tine Kleif and Wendy Faulkner, "'I'm No Athlete [but] I Can Make This Thing Dance!' Men's Pleasures in Technology," *Science, Technology & Human Values* 28 (2003): 296–325.

24. Carolyn Marvin, *When Old Technologies Were New* (New York: Oxford University Press, 1988), chap. 1; Sally Hacker, *"Doing it the Hard Way": Investigations of Gender and Technology* (Boston: Unwin Hyman, 1990).

25. Carroll Pursell, "The Rise and Fall of the Appropriate Technology Movement in the United States, 1965–1985." *Technology & Culture* 34 (1993): 630.

26. Ibid., 635.

27. See Katherine Chen, *Enabling Creative Chaos: The Organization behind the Burning Man Event* (Chicago: University of Chicago Press, 2009).

28. See Yochai Benkler, *The Wealth of Networks: How Production Networks Transform Markets and Freedom* (New Haven, Conn.: Yale University Press, 2006); Daniel Kreiss, Megan Finn, and Fred Turner, "The Limits of Peer Production: Some Reminders from Max Weber for the Network Society," *New Media &*

Society 13 (2011): 43–259; also Adam Fish, Luis Murillo, Lily Nguyen, Aaron Pan-ofsky, and Christopher Kelty, "Birds of the Internet," *Journal of Cultural Economy* 4 (2011): 157–87.

29. Kreiss et al., "The Limits of Peer Production."

30. Gabriella Coleman, "The Political Agnosticism of Free and Open Source Software and the Inadvertent Politics of Contrast," *Anthropological Quarterly* 77 (2004): footnote 10.

31. Henry Jenkins, *Convergence Culture* (New York: New York University Press, 2006).

32. Independent Media Centers, Anonymous, and Riseup are activist techni-cal projects that struggle with the politics of inclusion/exclusion of people with differing levels of expertise, whereas Tor and open source software projects are comprised more uniformly of technical experts, thus obviating some of the con-flicts between engineers and laypeople. The growing open source hardware move-ment may bear closer comparison to technological media activism than software projects. Adam Fish et al. warn against generalizing about peer production in "The Limits of Peer Production."

33. Barley and Kunda warn against what they call "conceptual inversion," arguing that the notion that "networks are not hierarchies" is overstated. Stephen Barley and Gideon Kunda, "Bringing Work Back In," *Organization Science* 12 (2001): 76–95.

34. Interview, July 25, 2006.

35. Thanks to Biella Coleman for her help in drawing out these points.

CHAPTER 6

Participatory Design and the Open Source Voice

Graham Pullin

Participation in Principle and in Practice

Participatory design places the people who might use whatever is being designed at the heart of the design process. In other words, it means designing with them, not just for them. This can lead to designs that are more appropriate, and interventions that are more likely to be appropriated.

Participation can be as much an ethical and political issue as it can be a practical issue. "Nothing about us without us" is a mantra rooted in disability rights—and what design does not, after all, involve disability or diversity?[1]

In augmentative and alternative communication (AAC), many people who cannot speak use speech-generating devices. Participatory design can help lead to communication devices that better fit into people's lives. One of the key principles underlying practice and research in this area is the participation of people who use AAC every day.[2]

However, even when participatory design is embraced in principle, it can remain challenging to put into practice. Participation inherently involves deep challenges, beyond the difficulties in self-expression and communication already faced by people who use augmentative communication. There are many aspects of augmentative communication that we would like to see driven by the participation of those who use it. However, we have found that there are often factors inhibiting this deeper engagement.

For example, in a completely unsuccessful attempt to elicit the needs and wishes of a group of people who use augmentative communication, we asked direct questions:

Imagine your own dream communication aid ... what does it look like? What does it do? Where does it go? What does it let you do? What does it sound like?

Despite the central involvement of well-connected and well-respected AAC user Alan Martin in distributing and championing the survey, the responses were disappointing—if not unexpected.

A typical desire articulated was for "a Liverpool accent." While a perfectly valid request, regional accents are already available, and more still are under development by the world's leading speech technology providers, such as Acapela and Cereproc. This is already a well-recognized wish, so such requests did not illuminate any new perspectives.

Another (not uncommon) retort was, "I'd like something that f***ing works!" In these cases, it appears that the day-to-day frustrations of living with unreliable communication devices made optimistic speculation about future development seem like a naïve or wasteful task, at least to many of the AAC users involved.

Cultural Probes and Voice Parcels

Engaging people—any people—in imaginatively envisioning future possibilities, without being overly constrained by the present, is never a trivial exercise. In design research in general, techniques have emerged to elicit creative responses from users who might be bemused by the idea of exploratory design projects and reluctant to contribute their opinions. The use of "cultural probes" is one respected technique, originally introduced by Bill Gaver, Tony Dunne, and Elena Pacenti in 1999.[3]

Cultural probes—or "design probes," as they are sometimes called—are so named because they are left with a community and return fragmentary data to designers over time. They are deliberately tangential and abstracted. The original and much-imitated pack of probes included postcards that people were asked to write to themselves in the future and post (back to the researchers).

In Gaver's words, "[Cultural probes] address a common dilemma in developing projects for unfamiliar groups. Understanding the local cultures was necessary so that our designs wouldn't seem irrelevant or arrogant, but we didn't want the groups to constrain our designs unduly by focusing on needs or desires they already understood. We wanted to lead a discussion with the groups toward unexpected ideas, but we didn't want to dominate it."[4]

No one has applied cultural probes to augmentative communication—a field in which even the use of personas is considered radical, and is just beginning to be explored.[5] Attempting to inspire radical yet grounded futures is a profound role for design research to play. Augmentative communication is a field that in turn could influence—if not revolutionize—more mainstream applications for speech technology.[6]

The use of cultural probes is central to *voice parcels,* an ongoing partnership with Paul Gault, who has himself worked with disabled people to visualize their feelings about their disabilities, raising awareness among designers in the process.

One component of *voice parcels* involves asking participants to represent their voices metaphorically, through materials. Materials offer a very powerful way of engaging with abstract qualities: All individuals develop more or less nuanced sensibilities for their aesthetic, tactile qualities, and also their cultural and personal associations. A simple, iconic form has been chosen to represent the voice itself: that of the old-fashioned loudhailer, used by early-twentieth-century politicians addressing a public rally or by rowing coaches encouraging their crews. Participants will be asked to make or define a cone (with a handle, provided) which features one material on the convex outer face and another on the concave inner face. A few materials—silk, sheet steel, sandpaper—will be included in the pack to seed ideas, but participants will be encouraged to source their own. Beyond what the individual materials convey, we believe that the contrast—or even contradiction—between the two materials might illuminate a deeper complexity about a person's personality or identity. We are excited to find out how this format will be appropriated by our participants.

Voice and Tone of Voice

Users of current communication aids find them frustratingly lacking in expressiveness. "I want to be able to sound sensitive or arrogant, assertive or humble, angry or happy, sarcastic or sincere, matter of fact, or suggestive and sexy," said Colin Portnuff, who had amyotrophic lateral sclerosis, a progressive neurodegenerative disease. However, the only way he was able to affect the way his own speech-generating device delivered a line was through punctuation: ending a sentence with either a period, a question

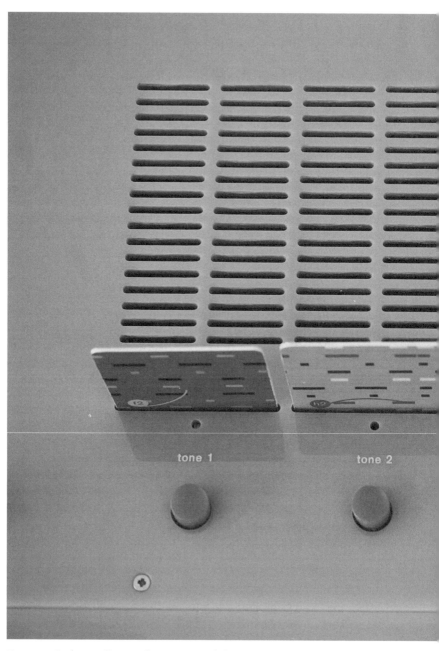

Figure 6.1. Graham Pullin, Marilia Ferreira, and Shannon Hennig, Tonetable, 2014; experience prototype with Moleskine™ notebook, RFID tags, and Arduino™ (with acknowledgment of Dieter Ram's T52 radio for Braun). Photo by Marilia Ferreira.

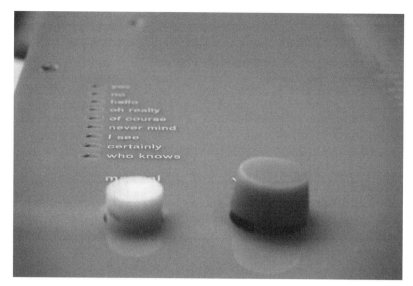

Figure 6.2. Detail of Tonetable showing selection of utterance. Photo by Marilia Ferreira.

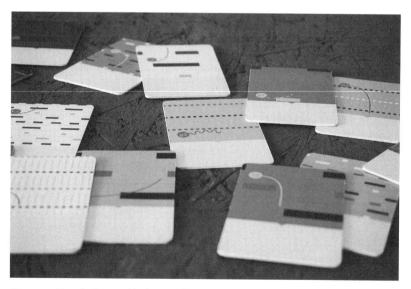

Figure 6.3. Detail of Tonetable showing deck of cards representing tones of voice. Photo by Marilia Ferreira.

Figure 6.4. Detail of Tonetable showing card inserted into slot. Photo by Marilia Ferreira.

Figure 6.5. Detail of Tonetable showing dual card slots for comparing two tones of voice. Photo by Marilia Ferreira.

mark, or an exclamation mark to change the way the words would be uttered by the text-to-speech technology inside.

Such a technique offers a very crude substitute for the nuances of spoken tone of voice. However, the technology is evolving; realistic yet more flexible speech synthesis is becoming technically possible—but we still need to define what it is that we want from this technology in everyday conversation. This is challenging territory even for phoneticians and linguists, who despite recourse to esoteric nomenclature find intonation and prosody elusive qualities, difficult to define and describe. In the context of participatory design, how on earth might we support a meaningful conversation among the rest of us? How might laypeople (as opposed to speech technologists) think of tone of voice in the first place, and thereby interact with it?

Three projects follow that explore, in different ways, issues pertaining to the tone of voice or expressiveness of speech-generating devices and how these issues might be opened up to people who use augmentative communication. At the same time, they explore how design—a traditionally visual discipline—might contribute to our understanding of the audible qualities of speech.

Experience Prototyping and Tonetable

As already mentioned, research in AAC needs the participation of people who use speech-generating devices. Given that most AAC research in this area is carried out by speech and language professionals rather than by speech technologists, the ability to experiment with the social role of tone of voice also needs to be opened up to those who don't have the technical skills to synthesize speech themselves.

Following the success of Six Speaking Chairs (discussed below) in engaging an AAC community, we came to believe that iconic research apparatuses could play a role in giving physical form to this otherwise invisible and overlooked research issue.

Tonetable, developed in partnership with Shannon Hennig and Marilia Ferreira, is currently at the stage of a pilot project. The apparatus would be capable of modifying the intonation of speech from any speech-generating device connected to it. Intentionally, it would not generate speech itself; the question of different voices and their associated identities ("I'd like a Liverpool accent") is a separate issue in the discourse surrounding speech-generating devices.

The physical design of Tonetable has been prototyped. It involves a deck of twenty-two cards, each one representing a different tone of voice. Participants can select a card and insert it into either of two card readers, allowing comparisons between two contrasting tones of voice in a particular conversational context. So as not to influence perceptions, the cards are unnamed except for an alphanumeric identifier. Abstract patterns aid recognition and are intended to support a growing conversation between researcher and participant about each tone. Each card also includes a blank area that can be written on, should participants wish to name or label a specific tone.

An "experience prototype" has been built, demonstrating interaction with the apparatus, although the final technology has not been implemented.[7] Therefore the apparatus has not yet been distributed, but Tonetable would also come with a notebook preformatted with tables in which experimental results and observations could be recorded. Researchers' notes would be handwritten, but could later be photographed or scanned and uploaded onto a communal research portal, where results could be shared and discussions hosted.

The participation we are aiming for is therefore twofold: between researchers and people who use augmentative communication, and also between different research centers. The latter feels important if the cultural nuances of tone of voice are to be recognized—and embraced. As an AAC user, Colin Portnuff lent authority to calls for participation of both kinds when he said, "Spend time with us. Learn from us, and teach us. Share what you learn freely and openly with your colleagues."[8] Making the investigative processes accessible to the entire community would catalyze new and deeper lines of inquiry.

Critical Design and Six Speaking Chairs

Six Speaking Chairs, conducted in partnership with Andrew Cook, was a research project intended to open a discussion about tone of voice between people without speech and people working in the field of augmentative communication.[9] As its name suggests, the core of the project was a collection of six chairs, each inspired by an academic or creative field that had an interesting view on tone of voice. Bringing all six views into a common medium, the chairs embodied these ways of thinking. Each chair incorporated a physical and interactive manifestation of one of the viewpoints,

Figure 6.6. Graham Pullin and Andrew Cook, Six Speaking Chairs, 2006–2010 (installation view, DJCAD, 2010); design collection with materials, interactions, and reclaimed chairs. Photo by Andrew Cook; copyright Pullin and Cook.

rendering it more engaging and more accessible to a lay audience than its original expert nomenclature.

The goal was thus to provoke new and inclusive discussions about tone of voice. The chairs were not communication devices in the sense already introduced, but rather conversation pieces: Their role was not to enable speaking itself but to enable speaking about speech. In this, the project could be understood as an example of *critical design*, a term defined by Anthony Dunne and Fiona Raby as design that asks questions rather than design that solves problems.[10] Dunne and Raby's pioneering project Placebo involved the creation of a collection of objects related to the invisible electromagnetic radiation all around us.[11] The adoption of these objects by the public, in their own homes, prompted reflection on the everyday implications of electromagnetic fields. It elicited rich nar-

ratives that illuminated deep preoccupations with health, privacy, and well-being.

Although the chairs were not intended strictly as early prototypes of new, functional communication devices, they were nonetheless interactive. They were "experience prototypes," which means that a variety of technologies were brought to bear, sometimes technologies very different than those that would be employed in an actual communication device.[12] What mattered was that they could be examined, sat on, tried, and discussed. Mounted beside the seat of each chair was a gray box containing a prototype user interface. Using the box, someone sitting on the chair could modulate the tone of voice of the synthetic speech generated by an old-fashioned horn speaker at the front of the chair.

The vocabulary was very restricted: to just four words, in fact. This made the prototyping easier, but the decision was primarily intended to reverse the imbalance of current communication aids: Rather than being

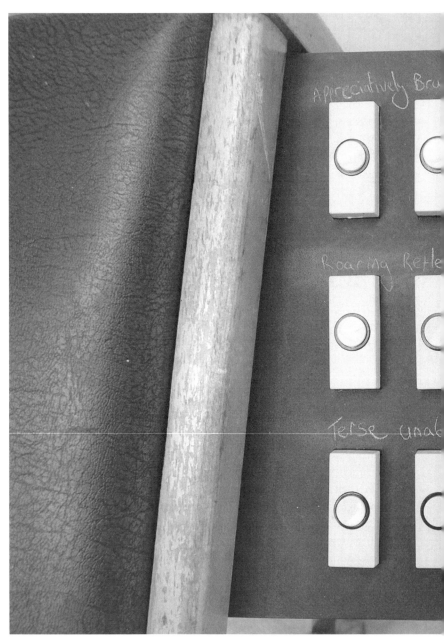

Figure 6.7. Chair No. 6, "The Terse/Roaring Chair," 2008; above each of the seventeen doorbells are handwritten descriptors taken from George Bernard Shaw's script for Pygmalion. *Photo by Graham Pullin; copyright Pullin and Cook.*

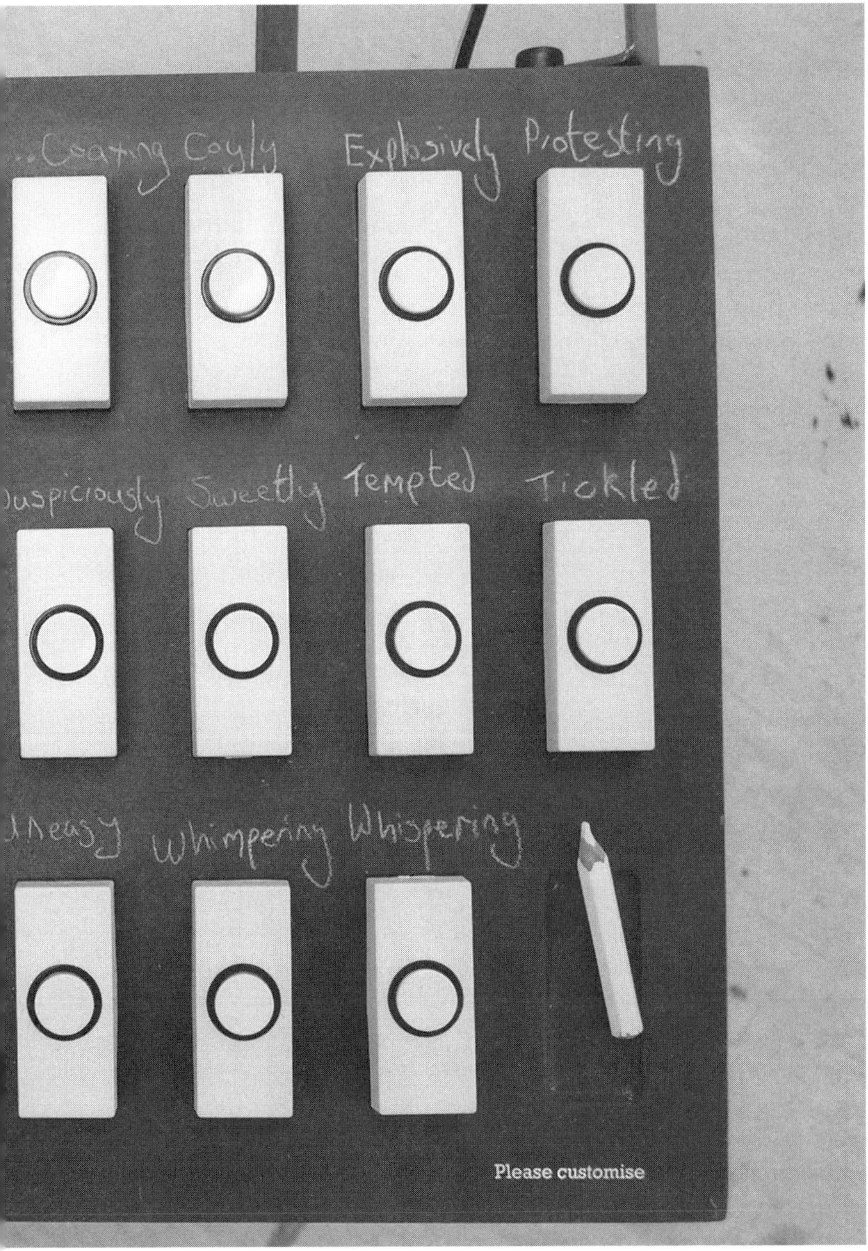

..Coaxing Coyly
Explosively
Protesting
uspiciously
Sweetly
Tempted
Tickled
neasy
Whimpering
Whispering
Please customise

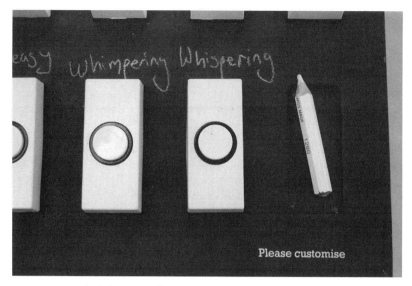

Figure 6.8. Detail of Chair No. 6, "The Terse/Roaring Chair," 2008; note the white pencil and the invitation to "Please customise." Photo by Graham Pullin; copyright Pullin and Cook.

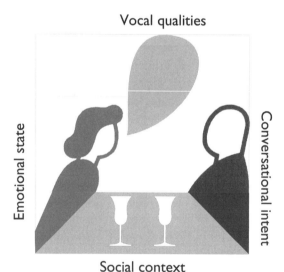

Figure 6.9. Four perspectives on describing tone of voice (with acknowledgment to Gerd Arntz's signs for the Isotype system); from Graham Pullin, 17 Ways to Say Yes, unpublished PhD thesis, University of Dundee, 2013.

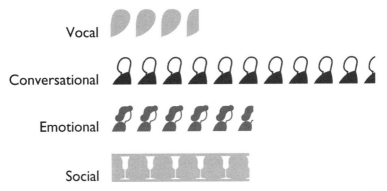

Vocal

Conversational

Emotional

Social

Figure 6.10. Mapping 257 tones against four perspectives—each icon represents ten of the 257 tones (with acknowledgment of Otto Neurath's Isotype system); from Graham Pullin, 17 Ways to Say Yes, *unpublished PhD thesis, University of Dundee, 2013.*

able to say many words in a very limited number of ways, these prototypes were capable of saying a few words in a variety of ways. Sitting on each chair and using a control to the left of its seat, a participant could select between "yes," "no," "really," and "hello." These words were chosen because their meanings can be radically altered depending on how they are said.

The chair that involved the deepest audience participation was Chair No. 6, "The Terse/Roaring Chair." It had seventeen doorbells, above each of which was scribbled a description of various speaking styles, including *appreciatively, brusquely, coaxing, coyly, explosively,* and *protesting* (these descriptions refer to stage directions in George Bernard Shaw's script for *Pygmalion*).[13] In the bottom right-hand corner, after the last two doorbells, marked *whimpering* and *whispering,* a short white pencil sat in a shallow trough, next to the words "please customize."

This invitation was used to frame a participatory exercise at ISAAC (the International Society of Augmentative and Alternative Communication) 2008 in Montréal with an audience that included speech and language professionals and people who use AAC.[14] Participants were introduced to the chairs and then asked to customize the sixth by writing up to seventeen additional ways to possibly say "yes," imagining themselves restricted to these seventeen for the rest of their lives. In the case of people already using limited speech-generating devices, seventeen different ways might exceed their current capabilities. The responses of forty individuals

Figure 6.11. Ryan McLeod and Graham Pullin, Speech Hedge, 2010; concept visualization for Toby Churchill's Lightwriter™ and Apple's iPhone™ (with acknowledgment of Orla Kiely's Multistem™ fabric). Photo by Ryan McLeod.

were collated. Even after combining equivalent terms (such as anger, angrily, and angry), this resulted in over 250 distinct descriptors.

When these descriptors were sorted into categories, four perspectives were identified: emotional state, conversational intent, social context (in terms of the speaker and listener's relationship, status, or social setting), and vocal qualities (whether described directly or metaphorically). The icons in the diagram are taken from Gerd Arntz's signs for the Isotype system.[15] Like the recycled chairs and the conceptual frameworks taken from various speech-related disciplines, these are yet more found objects.

When the 257 tones were mapped against these four perspectives, it was striking how few could be classed as descriptors of emotional states. This result challenges some assumptions at the heart of the vast majority of speech technology research and development: that an "emotional model" can be used to describe expressive speech and that any tone of voice can be interpolated between speech patterns corresponding to extreme happiness, sadness, anger, and fear. The addition of other perspectives implies that tone of voice is more complex than these assumptions suggest. Not that there is anything absolute about these perspectives; it is just that ordinary people, when they reach for descriptions of tones of

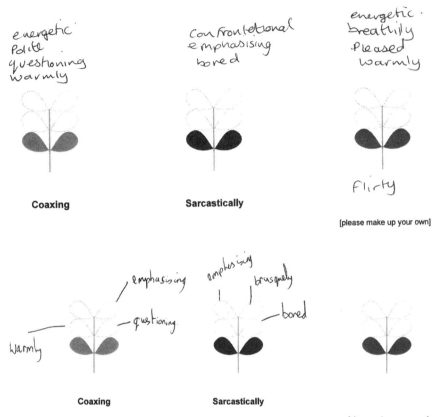

Figure 6.12. Two participants' responses from the Million Ways to Say Yes pilot study at ACE North, 2010; from Graham Pullin, 17 Ways to Say Yes, unpublished PhD thesis, University of Dundee, 2013.

voice, use a more heterogeneous collection of perspectives than academics do (academics are, after all, often engaged with deliberate attempts to propose unified or simplified models).

Speculative Design and Speech Hedge

How this tonal complexity might be reconciled with a simple and intuitive user interface was explored in a second project, in partnership with

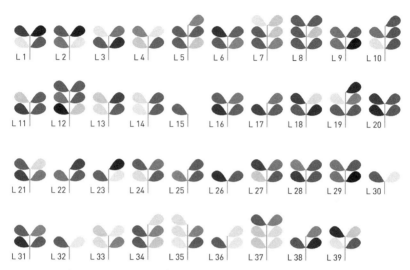

Figure 6.13. Thirty-nine participants' responses to "coaxing" from the Million Ways to Say Yes pilot study at ACE North, 2010 (with acknowledgment of Orla Kiely's Multistem™ fabric). The position of the individual leaves follows the participants' hand-drawn versions; the colors are taken from Speech Hedge, in which some colors have meaning (red for loudly, putty for bored, chocolate for rich voice, pale blue for breathily, orange for energetic) but others are chosen for a coherent overall color palette; from Graham Pullin, 17 Ways to Say Yes, unpublished PhD thesis, University of Dundee, 2013.

Ryan McLeod. Speech Hedge was a "design exploration"[16] or "speculative design"[17] that took the four lay perspectives on tone of voice as its starting point.[18]

It imagined that a standard text-to-speech communication device could be complemented by a user interface devoted to controlling tone of voice, perhaps running on a smartphone. In practice the devices could be integrated, but for an augmentative communication and speech technology audience, the separation made it very clear what functionality had been added.

People using the system had direct access to sixteen tones of voice. These tones were represented by little diagrammatic "plants," inspired by the Multistem fabric designs of Orla Kiely.[19] The collection of sixteen plants thus formed the "hedge" that gave the concept its name. Each individual plant was assembled from between two and eight "leaves," and each leaf represented an element of tone of voice (as selected from the

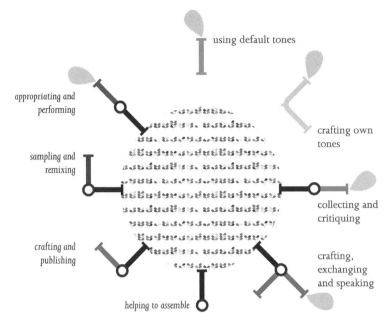

Figure 6.14. Diagram of a diverse community of people variously crafting, exchanging, browsing, critiquing, collecting and speaking (those indicated in italics might be speaking people); from Graham Pullin, 17 Ways to Say Yes, unpublished PhD thesis, University of Dundee, 2013.

diverse responses to Chair No. 6, our participatory exercise with the seventeen doorbells).

If this concept started off as an exercise in visualizing thousands (if not millions) of different tones of voice, it soon evolved into something more radical still.

Again the central concept was used to frame a participatory exercise: in this case, one that asked people to synthesize a given tone of voice from simpler elemental tones. Since the speech technology had not at this point been prototyped, participants were invited to draw metaphoric plants — composed from the various leaves and their corresponding properties — on paper.

Some responses further illuminated the different perspectives on tone of voice already discussed. One respondent aimed to synthesize *sarcastically* by combining *confrontationally* and *politely*, while another approached it by combining *emphasizing, loudly,* and *brusquely.* The first reads as a top-down description of the contradictory conversational and social role of sarcasm, while the second can be seen as a bottom-up description based solely on its sound. Other comparisons revealed a further effect: One respondent assembled *sarcastically* by combining *bored* and *emphasizing*, while another combined *loudly, energetic,* and *emphasizing*. The first reads as a deadpan delivery of sarcasm, the second as an ironically exaggerated delivery.

This time two quite different sounds were being described—and who is to say which of these definitions is most appropriate? What might be seen as a problem could, if embraced, catalyze a revolutionary approach to speech technology development. A feature of Speech Hedge is that once someone has created a tone of voice, they get to label it—subjectively. This not only makes the concept feasible (because there are no technical criteria for automatically specifying the precise social function of a tone of voice): It is also an ideal. A precedent for this form of subjective labeling within a communal collection can be found in another field. An Adobe software product called Color CC (previously called Kuler) enables a community of graphic designers to share, exchange, and critique one another's color palettes.[20] The designers label their color palettes with evocative names that can resonate within the wider community, seeding peer groups and subcultures.

The Open Source Voice

The question is therefore not just how tone of voice can be defined, but, more profoundly, *by whom?* An open source model is already common in speech technology, but it currently remains accessible only to a community of expert developers. What if speech technologies were conceived of as open source media in a deeper sense, in which myriad tones of voice could be crafted, exchanged, and appropriated by the very people who use them in their everyday lives?

This is a far more radical proposition, and one that challenges the notion that disability-related design should always benefit from—should always wait for—the so called "trickle down" of technologies from mainstream markets. Such apparently niche applications could, conversely, catalyze new developments for technology as a whole. Augmentative communication could yet be a crucible for radical advances in more expressive speech technologies.[21]

Notes

1. James Charlton, *Nothing about Us without Us* (Berkeley: University of California Press, 1998).

2. Sarah Blackstone, Michael Williams, and David Wilkins, "Key Principles Underlying Research and Practice in AAC," *Augmentative and Alternative Communication* 23 (2007): 191–203.

3. Bill Gaver, Tony Dunne, and Elena Pacenti, "Cultural Probes," *Interactions* 6, no. 1 (1999): 21–29.

4. Ibid., 22.

5. Jeff Higginbotham, Katrina Fulcher, Haesik Min, Carly Hanna, and Carrie-Anne Kirkland, "Developing Personas to Aid in AAC Design," paper presented at American Speech-Language-Hearing Association Convention, Philadelphia, November 18, 2010.

6. Graham Pullin, *Design Meets Disability* (Cambridge, Mass.: MIT Press, 2009).

7. Marion Buchenau and Jane Fulton Suri, "Experience Prototyping," paper presented at 3rd International Conference on Designing Interactive Systems, New York, August 17–19, 2000.

8. Colin Portnuff, "Augmentative and Alternative Communication: A User's Perspective," lecture delivered at the Oregon Health and Science University, August 18, 2006, http://aac-rerc.psu.edu/index-8121.php.html/.

9. Graham Pullin and Andrew Cook, "Six Speaking Chairs (Not Directly) for People Who Cannot Speak," *Interactions* 17, no. 5 (2010): 38–42.

10. Anthony Dunne and Fiona Raby, *Design Noir: The Secret Life of Electronic Objects* (London: August/Birkhäuser, 2001).

11. Ibid., 75.

12. Buchenau and Fulton Suri, "Experience Prototyping."

13. George Bernard Shaw, *Androcles and the Lion, Overruled, Pygmalion* (London: Constable, 1916).

14. Graham Pullin and Andrew Cook "Six Speaking Chairs to Provoke Discussion about Expressive AAC," paper presented at 13th Biennial Conference of International Society for Augmentative and Alternative Communication, Montréal, August 2–7, 2008.

15. Otto Neurath, *International Picture Language* (London: Kegan Paul, 1936; facsimile reprint Reading: Department of Typography & Graphic Communication, University of Reading, 1980).

16. Daniel Fallman, "The Interaction Design Research Triangle of Design Practice, Design Studies and Design Exploration," *Design Issues* 24, no. 3 (2008): 4–18.

17. Anthony Dunne and Fiona Raby, *Speculative Everything* (Cambridge, Mass.: MIT Press, 2014).

18. Ryan McLeod, *Speech Hedge,* video, https://vimeo.com/14582462/.

19. Orla Kiely, *Pattern* (London: Conran Octopus, 2010).

20. Adobe, "Color CC," https://color.adobe.com/sea-olf-color-theme-782171/.

21. Pullin, *Design Meets Disability,* xiii.

CHAPTER 7

Open Source Cancer

Brain Scans and the Rituality of Biodigital Data Sharing

Alessandro Delfanti and Salvatore Iaconesi

IN 2012, THE ITALIAN DESIGNER, open source activist, and digital media artist Salvatore Iaconesi open sourced his cancer. Inspired by a history of patient liberation and in an effort to demedicalize his condition, he placed all medical data and information related to his brain tumor— from brain scans to medical reports—on a website called La Cura (the cure)[1] alongside an open and inclusive request for "cures." Speaking through a YouTube video, Iaconesi explicitly addressed peers, physicians, activists, artists, designers, and engineers, asking them to engage with the data and to use it to produce any cure they could imagine for his condition.[2] He also promised to publish all of the cures so that others could use them. Iaconesi subsequently received and shared hundreds of thousands of contributions in forms as diverse as medical advice, artwork, peer support, and poetry. But what does it really mean to "open source" one's cancer? Like other illnesses, cancer can be regarded as a metaphor for different political and social orders.[3] Bearing this in mind, this chapter looks at collective action mediated by digital technologies in order to analyze a public and participatory experience of cancer.

The body is a battlefield where power relations are structured and negotiated. With the rise of pervasive interactive media, digital representations of bodies have come to have force in the material world. Thus the act of becoming a patient, to put it in Foucauldian terms, is changing, and new ways to resist (or reinvent) this process seem to be emerging. In an effort to explore this evolving landscape, we focus on the role of hacker cultures. "Hacker" is a polysemous term, one that encompasses several communities, practices, and subcultures. Hackers draw upon different political backgrounds while sharing an interest in craftiness, cleverness, and a refusal of bureaucracies in favor of decentralized solutions and organizational forms. Regardless of its political orientation and technical nuances, hacking is often concerned with performing technological alternatives in the public sphere in order to convey their emancipatory potential.[4] It is exactly by focusing on the symbolic significance of hacking biodigital data that we interpret the opening up of cancer's "source code" as a biopolitical rite of healing, aimed at redefining concepts such as "disease" and "cure." The act of accessing and sharing the medical data that forms this source code is symbolic of the desire to reappropriate the condition of being ill, and to foster a society that recognizes disease as a complex experience—one felt by social bodies as much as individual bodies. Open sourcing and crowdsourcing can be seen as dense biopolitical signifiers rather than mere distributed technical solutions. We also discuss the role played by a participant public in fostering these emergent understandings, a public mediated by digital platforms and gathered around the biodigital data shared therein.

This chapter is organized as a narrative. The temporal unfolding of La Cura is supplemented by insights and reflections garnered from digital culture studies, medical anthropology, and feminist theory. We should highlight that Iaconesi is himself one of the authors of this chapter, and thereby occupies a role as network-ethnographer embedded within the system being observed. Thus the material is culled from the direct experience of one of the authors, in addition to being based on repeated interactions and discussions with crucial individuals involved in the events, analyses of the media products related to La Cura, and a sample of the cures received. We would like to make the reader aware of two things: First, we have chosen to narrate in the third person to reflect the chapter's formation through long collaboration and conversation. Second, we interpret this chapter as one of the cures advocated by La Cura. A bit of background will help explain what this means. Now in his mid-thirties, Iaconesi is a well-known designer, artist, and open source advocate. He is a former TED Fellow, and, alongside his partner Oriana Persico, he cofounded and runs the Art Is Open Source collective.[5] Through La Cura, his experience of illness also became a media intervention seeking to expand the domain of what it means to confront cancer, moving beyond the state of medicalization and toward a scenario in which medical institutions are part of a broader system that includes one's social and affective worlds. The first step in this transformation was to convert his medical records from professional to common standards, making the data easily readable and shareable by laypersons. More abstractly, Iaconesi sought to construct an inclusive understanding of the word "cure," one that is not limited to a medical definition but that extends to different practices and meanings. La Cura had several aspects and implications. Extensively discussed in the public sphere—in both digital networks and major international media outlets—it also altered Iaconesi's relation to the professional medical sphere. For example, Iaconesi met the surgeon who ultimately removed his tumor as a direct result of the suggestions provided through the website.

Here we focus on La Cura's public character as a collective media intervention. To this end, we turn to traditional ethnographic accounts of healing rites and use them as a theoretical framework to propose La Cura as an example of a "biodigital ritual of sharing." This ritual follows a protocol or script derived from hacker practices and rhetoric, emphasizes public involvement, and opens cancer to a plurality of meanings and understandings. According to this script, one hacks into and expropriates data controlled by institutions, shares it in the open, and thereby facilitates the construction of a community around its free and unpredictable use.[6] Medical images are powerful signifiers of scientific authority,[7] and while for medical institutions a piece of data such as a brain scan represents an instrumental, objective abstraction from an individuated body, its symbolic reinscription through La Cura's biodigital ritual seeks instead to salvage the social body that medical institutions tend to discard. The ill person is not the sole or primary object of the rite. Instead, the ritual focuses on a broader target: attempting to fix social and political imbalances that affect the entire collective. From this perspective, we lend our efforts to an attempt at defining the significance of openness in relation to scientific and medical knowledge. Scholarly literature on biohacking, citizen science, and do-it-yourself biology insists on the role of openness, in terms of open access to data and tools, as a prerequisite for public participation in the "domestication" of biomedicine.[8] Although we acknowledge the importance of access to information, we add to the picture a consideration of the ritualistic and symbolic aspects of openness: La Cura framed access as a necessary but insufficient first step. It also sought to confront the symbolic apparatus represented by medical institutions and their processes, in order to imagine the possibility of plural degrees of freedom, diversity, and sociality. Still, the free circulation of digital artifacts is a prerequisite for the ritual, because it allows for the creation of a public that shares a common knowledge upon which a sense of participation can be encouraged and built.

While this specific case study is related to hacker cultures, we believe it offers broader insights into the biopolitics of health and illness as performed in the digital sphere. It also allows us to supplement understandings of contemporary biomedicine with other aspects of today's information environment. Institutions act as enablers and facilitators of digital participatory practices, yet at the same time they can be challenged by practices that happen outside of their control. As institutions typically attempt to incorporate and transform these practices, institutional and technical protocols need at times to be circumvented or cracked open. When this occurs, constraints, intentions, and conflicts can be revealed and confronted. Following feminist theory, we also suggest that, in the face of illness and disability, digital cultures often imagine and perform technologies as social and relational prostheses, as opposed to bodily prostheses.[9] The prosthetic role of technologies can, in fact, be focused on enhancing one's social capabilities as well as intervening in the materiality of one's body. Finally, in the conclusion we suggest that this example is part of an emergent culture of *digital solidarity,* wherein the building of common, open, and autonomous spaces is advocated as an alternative to institutional crises of legitimation.[10] By simultaneously acknowledging the risks of such endeavors, we suggest that they can pave the way for new forms of medicalization, surveillance, and health-care privatization.

From Scars to Scans: Open Source as a Symbol

When you have cancer you disappear, only to be replaced by something else: a patient. The patient is a strange being that is, on the one hand, entirely made of data: blood exams, images of body parts, lab values, diagnoses—the list goes on. On the other hand, this data is not simply *you:* it barely approximates the full complexity of selfhood. As a patient you are suspended from the world around you. This is a major transformation, in which all daily activities and routines tend to cease as the patient enters into a crisis. You don't eat the same anymore, you don't work at the same pace, you don't spend your time in the same ways or with the same people. New routines, schedules, bureaucracies, administrative tasks, and people appear and surround you. All these new things in your life relate to the disease. Almost everything else is erased. These processes are bolstered by the fact that you are also objectified through a set of data. Your body, personality, and social connections disappear, and are replaced by data and images. Everyone around you begins speaking in terms of these data and images. Your prior self vanishes as you are taken over by the disease, on one side, and by the data and images, on the other. Language changes: different words, pauses, timings, embarrassments, taboos, phrasings, and different ways of saying—or not saying—things. People cease to simply talk *with* you; rather, they speak *about* you, and only discuss the patient that you have become. In August 2012, in Rome, after several episodes of epilepsy, Iaconesi was diagnosed with a brain tumor. One day, in the hospital following several brain scans including CT (computerized tomography) and MRI (magnetic resonance imaging), he asked for a digital image of his cancer. He wanted to look at it, to see this thing that was growing inside of him. But his request was denied. Everything got in the way: from administrative red tape to privacy and legal issues, and barriers imposed for insurance reasons. Everything about the medical system made it impossible for Iaconesi to access his own cancer. He had the distinct feeling that this situation was not about him, but about a medicalized version of himself, a self reduced in complexity. This situation triggered his first response: After several attempts at consultations with doctors and surgeons, Iaconesi left the hospital against their advice. At that time, nobody knew his tumor was a benign and removable glioma, and the specter of death was still implicit in the diagnosis. On leaving the hospital he signed an agreement lifting all responsibility from the hospital, and requested his medical records, including brain scans, in digital format on a DVD. Only by assuming full responsibility for his condition was he able to put in motion the bureaucratic machinery that would provide him with the images. He finally had the files but, upon arriving home, he discovered to his chagrin that while the files were in a format that is technically open (DICOM), they were not suited for access and use by nonprofessionals.[11] One needs specialized software to open these files, and even if one can open them, they are not really meant for a layperson to use: The language, terminology, and nature of the icons are completely abstracted from the layperson's experience. The images of Iaconesi's cancer were only meant for technicians and physicians. Iaconesi felt it impossible to easily translate such data into his complex social world, which includes global digital relationships built through his activity with the Art Is Open Source collective.

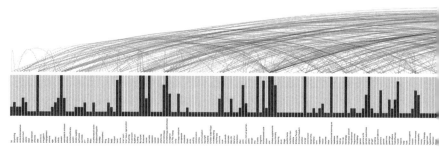

Figure 7.1. "Topics and Relations," the information visualization about the topics discussed on La Cura and their relations. By Salvatore Iaconesi. Image courtesy of the artist.

Modern processes of medicalization tend to extract and isolate the ill person from her social, cultural, and perceptive contexts. In the medicalized sphere, disease is no longer perceived as a social matter or an object of shared, societal action. Instead, it is a specialized object reserved for treatment by professionals and institutions. In a biomedical model of illness, ill health is in fact a "deviation from . . . the normal range of measurable biological variables."[12] Imaging technologies such as X-rays or PET scans are just one vector along which the medical gaze objectifies their use, privileging images and data over individual experience and, crucially, even the body itself.[13] The increasingly pervasive digitization of life (like that which occurs in genetics, for example) has taken this phenomenon further still, pushing toward a reconceptualization of the body as pure information. This dematerialization accounts for a *disappearance* of the individual and her complexity.[14] The brain scans generated in the course of Iaconesi's treatment were restricted to professional settings and were thus central in defining the strict boundaries of his role within the hospital. He felt that within medical institutions his affective and social relationships were being rendered invisible, as if his body was being separated from his social world. This had multiple manifestations, including the modulation of his experiences of time and space in the hospital: the strict wake up calls; the fixed routines for doctor's visits and the assumption of drugs, meals, and visits from relatives and friends; and constraints on movement within and outside the hospital. However, the loss that he felt most deeply was the progressive disappearance of those activities and relationships that enrich one's life: from music, art, culture, hobbies, and preferred food, to the negation of friends' and relatives' participation in activities such as

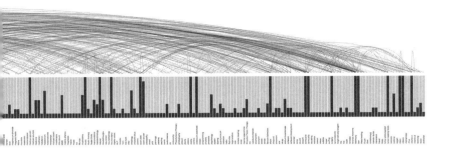

cooking or conversation. For medicalized individuals, this "pathway from person to patient"[15] is a common experience that coincides with a heightened vulnerability to biopower. Iaconesi's case clearly bore out this analysis, but the reaction that led to La Cura was primarily triggered by a recognition of how biodigital information was produced, inscribed in digital supports, and shared. Foucault insisted that technologies of governance are composed of practices and objects through which biopower is exerted and resistance emerges.[16] Iaconesi's brain scans quickly assumed this dual role and became a symbol of different social orders: first signifying medicalization and only subsequently facilitating processes of individual *reappearance* and collective reappropriation.

As soon as he found out the file format of his medical records, Iaconesi worked to convert the files into more ordinary, shareable formats like JPEG, DOC, and HTML. To do so, he had to write his own software code. It was at this moment, sitting intently at his computer — having left the hospital and returned home determined to share his medical records — that those around him again recognized him as the Iaconesi they had known. He reappeared. He ceased being a patient and became a human being again. Hacking and sharing his medical records was the most natural thing he could do to reconnect with his broader social and affective world. Only days after he received his diagnosis, according to his partner Oriana, hacking the data allowed them to regain control over the dispossession experienced at the hospital. While the DICOM standard embodied the biopower experienced in the hospital — reinforcing the disappearance of the person in favor of the patient — by reinscribing the data through an open source approach, Iaconesi effectively reappropriated the medicalized experience of cancer. Biodigital data assumed the "polarization of meaning" that Turner describes as inherent to the symbols used in rituals, becoming simultaneously associable with a physiological fact (cancer) and a social fact (resistance to objectification).[17] Iaconesi set up a website with his medical data available for download by anyone: brain scans, blood tests, medical reports, and diagnoses—all in easy-to-share formats. Through data sharing, he resisted his reduction to a mere cancerous body that was composed only of a set of medical data.

But what does "hacking" mean in this context? Among the many facets of hacking, here we focus on its communicative and performative side.[18] The media interventions constituted through data hacking and sharing are aimed at the "transformation of suffering into communicative signs," as Tamar Tembeck has noted in the context of visual autopathographic practices. Hacking subverted the standard cultural meaning of what being a "good patient" means, including the acceptance of an objectified role in the machinery of medicalization.[19] In converting his medical data to more widely accessible digital formats, Iaconesi circumvented two codes. The first code was plainly visible: the digital code that underwrote the brain scans themselves, making them legible only to medical professionals. The second code was rendered visible by the hack: It was the institutional code that maintained the specific social order found in modern health systems, inscribed into the DICOM format, with the effect of preventing Iaconesi from using the scans in a way not envisioned by medical institutions. These codes were easily representable as the legacy of calcified bureaucratic systems that needed the infusion of new practices and values. In contemporary societies openness is more than just an organizational principle. Free access to information can symbolize concurrent values, like the desire to resist privatization or democratize the political sphere.[20] Performances related to transparency and openness have been used in countercultural activist approaches to cancer since the 1990s. For example, by publicly displaying one's body after a mastectomy, some women have famously reclaimed the scar as an "object of political significance."[21] While Iaconesi didn't have a scar (yet), his brain scans alone became objects of political significance in their exposure. Yet this was not related to the need to overcome the social stigma associated with cancer. In La Cura, openness was used to contrast the ways in which a person is transformed into a patient in the medical system. It symbolized the possibility of reverse-engineering this transformation by reappropriating the abstracting patient data and inserting it into a wider process: a process that included one's social, political, relational, economic, affective, and creative life, as well as the willingness to engage other members of society. In this performance of hacking, open source data exposed Iaconesi's fight against a form of medicalization exacerbated by digital technologies. This was a performance in more than one sense. First, hacking the data allowed Iaconesi to construct and reclaim an identity that was not reduced merely to that of a "patient"; it was an act of writing oneself into being, as suggested by Jenny Sunden's explication of self-reflexivity based on the materiality of textual bodies constructed through digital media.[22] Additionally, open sourcing the data allowed for the creation of a public that performed and participated in the biodigital ritual hinted at above and discussed further in the next section. La Cura became a platform for performance through shared and recombinant reproduction, reappropriation, and reinterpretation. The result of the initial action was the production of a performative space across a variety of media, and the inaccessibility of Iaconesi's data set the scene for a translation of hacker rites into the world of cancer.

La Cura: Participation in the Ritual

On September 10, 2012, Iaconesi posted two videos on YouTube, one in Italian and one in English. The video, entitled "My Open Source Cure," begins with a simple declaration: "I have brain cancer." Iaconesi continues:

> This is a cure. This is my open source cure. . . . In different cultures the word cure means several different things. There are cures for the body, cures for the soul, there are cures for communication, for socialization. So what I ask you is to give me a cure. . . . Use the data and information in open format that I published to produce something . . . to produce a video, a graphic, an artwork, a game, or maybe even study the information to find a cure for me. . . . I will publish all the cures that you send me so that everyone will be able to benefit from them.[23]

The video sought to enlarge the scope of what a cure is or can be. Iaconesi invited everyone to join him in his disease, turning it into "a matter of engagement for the whole society." He was interested in finding out how others could participate in his cure: What could an artist do with the data? What cures could be envisioned by a designer, an academic, or a hacker? The video went viral. Italian and international news outlets published the news that Iaconesi was asking people to help him find a solution to his brain cancer, and he was interviewed frequently.[24] Over the following months, he continued to post every medical record related to his condition: postsurgery reports, brain scans, blood exams, and histological results, in addition to other messages.

In the beginning, Iaconesi mobilized a network of artists and designers who had previously been familiar with his work, but it wasn't long before La Cura swelled outside of this network. Then the cures started pouring in. He began receiving and exchanging myriad messages, of all sorts and from multiple sources, both online and offline—through e-mails and social networks, from people in the streets and those on their phones, through live, physical visits from friends and strangers, and across continents through Skype and Google Hangouts. One medium would lead into another, as Iaconesi would respond to an e-mail by meeting its sender in person. The cures were as diverse as one can imagine. People contributed stories, artworks, medical advice, consultations, traditional healing, magic, spirituality, dietary advice, and offers of financial support. Cancer patients asked their own doctors for suggestions about Iaconesi's condition. Contributions were generated across the globe, with epicenters emerging in Italy, Spain, Slovenia, Croatia, Greece, and the United States, complementing numerous cures from Northern Europe, Chile, and Brazil. Most of the countries of the world were represented. In fact, it would be impossible to analyze the cures without using automated tools: Ultimately they comprised about one million texts (as of June 2014)—a number that does not include the Facebook and YouTube comments, tweets, e-mails, and face-to-face interactions that are still accumulating as we write this chapter.[25] Here we can only hint at some of the radically different cures received through the website. One of Iaconesi's old companions, whom he had not heard from since the latter had moved to Argentina several years earlier, advised him to visit Argentina, relax, eat natural food, and breathe clean air. A communication sciences professor suggested, in a beautiful, poetic way, a meeting with one surgeon in particular—the one, in fact, who would eventually perform Iaconesi's operation. Through social network interactions, e-mails, and phone calls, she related her own experience with the surgeon; she herself had been diagnosed with a meningioma, and she interwove a description of the doctor's treatment techniques with scientific evidence, a consideration of his Mediterranean warmth, and expressions of her personal feelings and emotions. Someone suggested a method

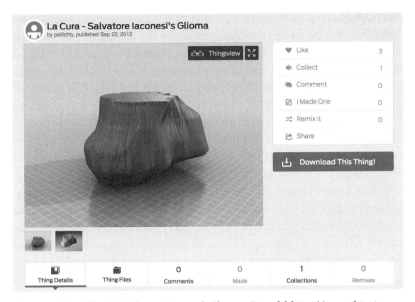

Figure 7.2. Patrick Licthy, Salvatore Iaconesi's Glioma, 3D model, http://www.thingiverse
.com/thing:30987/.

invented by a Brazilian monk: a mix of aloe vera, honey, and whiskey. A
person commenting on the initial La Cura video suggested that Iaconesi
might consider giving up his quest for a cure and simply kill himself
instead—a decision that would supposedly generate enormous amounts
of insurance money for his friends and relatives, securing their happiness
in the process. An artist created a performance that ritualized the magnetic aspect of MRI scans.[26] Another artist offered one of Iaconesi's favorite cures: Using the brain scans, he constructed a 3D design of the tumor
and made it available on Thingiverse, a website for digital design file sharing, so that anyone could print it out using a 3D printer and produce a
pocket-sized physical realization of Iaconesi's tumor (see Figure 7.2).[27]
Now his tumor could be with anybody.

When media outlets first became interested in La Cura, however, this
wealth of cures—and its tacit demonstration of the platform's inclusivity—
was still struggling to emerge. No matter how Iaconesi and Persico struggled to convey different images, the dominant narrative seized upon by
the mainstream media remained focused on the technicalities of La
Cura's open source, crowdsourced process.[28] This narrative performed

the technological determinist view that digitally distributed creativity is a technical solution. In this view, Iaconesi's open file sharing was presented as being narrowly concerned with facilitating a collective intelligence toward the discovery of a faster and more efficient technical solution to cancer. But, in our view, rather than being a positivist platform that advanced narrow conceptions of medical, technical progress, La Cura was more of a biopolitical performance that exposed the underlying power structures crystallized in the medical system. La Cura presented opportunities to reappropriate health and illness from the narrow, biomedical model of what being a patient means. As Arthur Frank notes, illness can provide justification for new desires.[29] Iaconesi's desires played a key role in the revelation of unexpected ways the codes at work in the medical world might be cracked—codes that barred patients access to data and produced an asymmetric relation between experts and nonexperts. Iaconesi thereby helped reveal and exploit the frictions and tensions inherent in the medical world.

Patients and their families turn to the Internet to retrieve information, discuss pathologies and treatments, provide peer support, share their stories or medical experiences with other patients, and mobilize.[30] Yet debates regarding patient inclusion and participation must be updated to include new forms of affective and social relationships that are changing the participatory landscape.[31] Digital networks can be seen as new "opportunity structures" that offer individuals the opportunity to participate in health management, while also potentially constraining and shaping their ability to act in other venues.[32] In fact, diffused visions of the "revolutionary" characteristics of digital media often overlook power dynamics, hierarchies, and failures that are integral to the Web. Rather than change power distribution patterns, spaces of digital media can reinforce and reproduce them—and even introduce new ones.[33] Authors such as Kate O'Riordan or Marina Levina, for example, respectively describe the emergence of *biodigital publics* and *digital networked subjectivities* as new forms of biopower based on the circulation of biological information through digital artifacts. For O'Riordan and Levina, the free circulation of biodigital artifacts enables collective public action toward reelaborations and reappropriations of health practices, but it also underwrites new forms of governance and capital accumulation.[34]

Building upon these arguments, our concept of a *biodigital ritual of sharing* allows us to account for other aspects of biodigital publics and chart the historical influence of hacker cultures on current practices of data sharing. Also, the concept acknowledges the rituality inherent to processes of digital community formation. Finally, instead of focusing on practices of sharing that are organized by commercial entities, such as companies that provide direct-to-consumer genetic testing and which themselves benefit from the digital labor provided by participant consumers, in this chapter we consider a public organized around an independent, digital commons. We believe that the reduction of biodigital sharing to a form of labor precludes a complete understanding of the complex entanglement of health and digital cultures. Public rituals are crucial symbolic and communicative acts that restate and renew basic social values and maintain community cohesion. Rituals can have a strong creative power, as they continuously recreate both the categories through which societies are perceived and constructed, and their underlying moral and relational orders. Rituals are standardized and divided into specific parts: They are scripts or protocols.[35] Furthermore, rituals related to healing, health, and illness are important parts of our experience and contribute to our social identities. In exploring the symbolic and ritualistic side of open source, we supplement traditional anthropological accounts of the social aspects of illness and healing with emergent, technologically mediated experiences. Processes of healing are intimately linked with the social structure in which the healing takes place. For example, according to V. W. Turner, the "rites of affliction" performed by the Ndembu people of Zambia symbolize that which "poisons" group life, demonstrating how rites of healing can perform a social regulatory function.[36] Indeed, rituals are meant to domesticate illnesses and resolve social conflicts, and are flexible enough to encompass new needs and adapt to change. In the case of La Cura, such flexibility was enabled by its public and participatory characteristics, which we regard as key to the emergence and stabilization of La Cura's digital public. The formation of a participant public is recursively key to the success of the ritual.[37]

By using digital technologies, participants of La Cura were able to ritualize their experience into a collective performance that built community. Indeed, drawing upon the notion of ritual, we suggest that desire was able to assert itself through a rite that aimed at remanifesting the social self previously negated by the medical institution. Via public rituals, symbols attain status as objects that mediate the reproduction and maintenance of social order, functioning as "restatements" of the terms of social life and human interaction.[38] A ritual involves specific objects and gestures in composing its choreography or script. The ritual of La Cura borrowed its choreography from hacker cultures. The public ritual of sharing, which is performed when hackers "liberate" information, follows a script that we can (over)simplify as follows: Data are hidden by bureaucracies using passwords, security, closed formats, and secrecy; somebody hacks into the data and shares it freely on the Internet; and an open community then makes unpredictable use of it.[39] Over the last few years, this script has become a crucial, and common, narrative within information societies—one need only consider the protest hackers Anonymous, the leaking organization WikiLeaks, or ex-NSA-contractor-turned-mega-leaker Edward Snowden to ascertain its significance. In the case of La Cura, data were hidden by the hospital, and access to it was denied to anyone who did not belong to the bureaucracy which deemed itself responsible for its management. The hospital denied Iaconesi access to his medical records, and when he finally received the files, they came in a format that did not allow him to use them. Iaconesi could not simply share the files: He had to hack into them and convert them into shareable formats. Finally, through La Cura he staged a media intervention that rendered the data visible, and aggregated a public that could interact with and use it in ways that were not controlled or normalized by medical institutions.

La Cura is part of broader countercultural approaches to health and illness. A direct inspiration for it was the history of patient liberation in 1970s Italian asylums. Psychiatrist Franco Basaglia, influenced by Foucault and the post-1968 social movements, set up open communication platforms (especially in the form of political assemblies) in which secluded mental health patients, physicians, and nurses shared desires, needs, and approaches to suffering. They reclaimed the will to understand, subvert, and eventually tear down the various walls that formed the fictional territory of hospitals, separating psychiatric institutions from the rest of society. Basaglia wanted to open up the physical, political, and knowledge-based power architecture of the institution. Besides changing how cures were performed, he imagined more open, inclusive, and permeable spaces and communities in which collective responsibilities were shared.[40] In her book on self-help feminist movements in 1970s California, Michelle Murphy describes what she calls "protocol feminism."[41] Protocol feminist groups would construct a procedural script that allowed for the spreading and maintenance of practices that they wished to encourage, such as collective self-examination outside of medical institutions. According to Murphy, these highly politicized scripts depended on cheap and accessible communication technologies and infrastructures—like photocopiers and highways—for dispersion. Feminist protocols of self-help were designed to reproduce and maintain a specific order of collective care and counterpower (what was once called *consciousness*). These practices were aimed at "seizing the means of reproduction," as Murphy puts it. In a fashion similar to the protocols described by Murphy, based on opening up one's own body to a sister's gaze, we suggest that biodigital rituals of medical data sharing can align participants with each other and maintain a social order that is currently imbalanced by a formalized institutional dominance. By collapsing the symbolic power of biodigital data and the choreography of hacking into a single practice, La Cura both created a social body that made the ritual possible and expressed the need for individuals to reappropriate experiences of cancer from medicalization.[42] The public visibility achieved through social and mass media was key in the building of a participant public. According to Lévi-Strauss, a participant group functions as a "gravitational field" for the ritual: It is key to its success and enables it to assume a significance that exceeds the role of the individual subject of the rite by expanding its relevance to the whole participant community.[43] The ill person is not the real object of the rite: In fact, Iaconesi's name was never mentioned in La Cura, and the website provided little personal information about him. At the center of the rite is a public concerned with social, collective imbalances. Utilizing biodigital rituals, a participant public can form when people identify with digital artifacts that are allowed to circulate freely in the mediated public sphere. This public can contribute to a ritual or ignore it and choose not to participate. The digital artifacts ultimately allow for the constitution of a mediated participatory form, capable of expressing the need to overcome societal frictions, obstacles, or hierarchies.[44]

Conclusions

In the wake of La Cura, Iaconesi feels that he has, in many ways, been successfully cured. The rituals of medicalization that articulate life in the hospital were shaken, and the institution acknowledged both changes triggered by the circulation of data in new spaces and Iaconesi's repositioning as a different kind of node within the information flux. Indeed, the international visibility and support he received from an ample community—

previously nonexistent—was accepted by and transformed his relationship with the medical system, as well as with his family and friends. His surgeons and his Chinese traditional healer contributed equally to aspects of the therapy he ultimately followed within health institutions. In the hospital, staff started referring to him as "Salvatore and Oriana," recognizing the significance of importing his affective world into the care they provided. Eventually his glioma was removed, and the choice of both the surgeon and surgical technique stemmed directly from La Cura and the relations that were established during the process. For example, while two surgical techniques had initially been identified as being equally effective, Iaconesi was able to incorporate in his choice a plurality of views that went beyond the usual technical considerations to include social, philosophical, and political considerations not commonly offered to patients. The method used involved the implantation of brain sensors, allowing the surgeon to perform functional tests. The final surgery was preceded by a critical conversation between the doctor and the patient, based on the results of those tests. Iaconesi was thus able to consciously take part in the decision processes connected to the surgery and its potential risks and effects.

La Cura thus enabled discussions about several important stakes. And more discussions seem poised on the horizon. The case of La Cura could, for example, be used to interrogate issues such as the role of digital media in patient empowerment, emerging forms of surveillance and resistance to them, the changing nature of medical expertise and knowledge, the construction of the body-self through digital data, and the role of the market in recuperating and managing such practices. This last issue is particularly important in La Cura. As the website became internationally visible, due in part to Iaconesi's appointment as a TED Fellow, Web companies interested in developing commercial applications for medical data sharing approached Iaconesi for advice. What kinds of participation and governance might stem from corporate services that organize patient biodigital information through open source approaches? The shift toward a more participatory, self-responsible, and proactive citizenship fostered by the digital sphere is related to similar transformations at the biopolitical level, where discourses of empowerment, participation, and collaboration constitute an emerging form of biopolitical governance. However, even as digitalization changes public views and practices, medical institutions can be ill equipped to respond to requests for radical inclusion, and often need to renew themselves by incorporating external practices. For example, pharmaceutical companies can emulate websites used by patient associations in order to establish a "sense of community,"[45] and can present Web services as seemingly neutral and democratic.[46] Indeed, hackers are entwined in cycles of incorporation and recuperation and often try to resist the way institutions adopt grassroots innovations to meet their needs.[47] Biodigital rituals of sharing in the context of interactive technologies might themselves become part of such a trend. Governmental medical power might easily extend to these emergent digital spheres, mixing self-control, responsibility, and radical transparency into a pervasive medical gaze.

While patient reclamation of the medicalized body is becoming a more common subject of discussion, by proposing the concept of the ritual we have here focused on the cultural significance of biodigital data: Once liberated through hacking from their objectifying role in medical institutions, open source data provide a commons upon which new forms of digital solidarity can emerge.[48] In doing so, such data can then trigger public responses that will enable collective reappropriations of standard experiences of cancer and other illnesses. Against techno-determinist ideologies, we also suggest that, by performing similar rituals of reappropriation and sharing, hackers and other members of digital countercultures can appeal to digital technologies, in addition to their bodies, as battlegrounds for the reconfiguration of social and political possibilities. In fact, hackers' technological and communicative skills can be used to construct spaces where power is called into question. A broader analysis of such rituals will need to explore the different cultural, technological, and political variables that shape forms of digital participation[49]—or nonparticipation[50]—as well as the way in which different pathologies can generate different forms of online organization and patterns of digital solidarity. Ultimately, La Cura signals the presence of a social imbalance that can be corrected through rituals that provide rallying symbols and facilitate collective interaction. This leaves us with the task of imagining other public rituals based on biodigital data sharing that might fix other social and political imbalances without exacerbating the present asymmetry of (bio)power, or accidentally introducing new asymmetries of their own.

Notes

We would like to thank Oriana Persico, a central participant in La Cura, as well as the thousands of people from all over the world who have joined Iaconesi in imagining a novel social and relational way of caring with and for people dealing with disease. We are also grateful to Gabriella Coleman, Tamar Tembeck, and the rest of the Media@McGill staff for their invaluable support and advice in the process of envisioning, discussing, and strengthening the ideas proposed in this text.

1. "La Cura," http://www.opensourcecureforcancer.com/.

2. Salvatore Iaconesi, *My Open Source Cure,* YouTube, http://youtu.be /5ESWiBYdiNo/.

3. Susan Sontag, *Illness as Metaphor* (New York: Macmillan, 1988).

4. For a rich account of hacker cultures and aesthetics, see Gabriella Coleman, *Coding Freedom: The Ethics and Aesthetics of Hacking* (Princeton, N.J.: Princeton University Press, 2013).

5. "Art Is Open Source," www.artisopensource.net.

6. See Gabriella Coleman, *Coding Freedom*; and Steven Levy, *Hackers: Heroes of the Computer Revolution* (New York: Delta, 1984).

7. Joseph Dumit, *Picturing Personhood: Brain Scans and Biomedical Identity* (Princeton, N.J.: Princeton University Press, 2004).

8. Morgan Meyer, "Domesticating and Democratizing Science: A Geography of Do-It-Yourself Biology," *Journal of Material Culture* 18, no. 3 (2013); Barbara Prainsack, "Understanding Participation: the 'Citizen Science' of Genetics," in *Genetics as Social Practice,* ed. Barbara Prainsack et al. (Farnham: Ashgate, 2014).

9. Donna Haraway, *Modest_Witness@Second_Millennium. FemaleMan©_ Meets_OncoMouse™: Feminism and Technoscience* (New York: Routledge, 1997); Michelle Murphy, *Seizing the Means of Reproduction: Entanglements of Feminism, Health, and Technoscience* (Durham, N.C.: Duke University Press, 2012).

10. Felix Stalder, *Digital Solidarity* (Lüneburg: PML Books, 2013).

11. "Digital Imaging and Communications in Medicine," Wikipedia, http:// en.wikipedia.org/wiki/DICOM/.

12. Mildred Blaxter, *Health* (London: Polity Press, 2010), 13.

13. Lisa Cartwright, *Screening the Body: Tracing Medicine's Visual Culture* (Minneapolis: University of Minnesota Press, 1995); Joseph Dumit, *Picturing Personhood.*

14. Margarete Sandelowski, "Visible Humans, Vanishing Bodies, and Virtual Nursing: Complications of Life, Presence, Place, and Identity," *Advances in Nursing Science* 24, no. 3 (2002).

15. Irving Kenneth Zola, "Pathways to the Doctor: From Person to Patient," *Social Science & Medicine* 7, no. 9 (1967).

16. Michel Foucault, *The Use of Pleasure: The History of Sexuality,* Vol. 2 (New York: Pantheon, 1985).

17. Victor Witter Turner, "Symbols in African Ritual," *Science* 179, no. 4978 (1973).

18. Coleman, *Coding Freedom.*

19. Tamar Tembeck, "Exposed Wounds: The Photographic Autopathographies of Hannah Wilke and Jo Spence," *RACAR: Revue d'Art Canadienne/Canadian Art Review* 33, nos. 1–2 (2008): 90; see also pages 94–96.

20. For an example of such a symbolic role see Endre Dányi, "Xerox Project: Photocopy Machines as a Metaphor for an 'Open Society,'" *The Information Society* 22, no. 2 (2006).

21. Lisa Cartwright, "Community and the Public Body in Breast Cancer Media Activism," *Cultural Studies* 12, no. 2 (1998): 117–38; see also Jean Dykstra, "Putting Herself in the Picture: Autobiographical Images of Illness and the Body," *Afterimage* 23, no. 2 (1995).

22. Jenny Sundén, *Material Virtualities* (New York: Peter Lang Publishing, 2003); see also David Gauntlett, *Media, Gender and Identity: An Introduction* (London: Routledge, 2002).

23. Salvatore Iaconesi, "My Open Source Cure."

24. La Cura was featured in media outlets such as BBC, CNN, *La Repubblica*, and *Le Monde*.

25. See http://www.opensourcecureforcancer.com/ for an initial analysis of the content, connections, and origin of the cures.

26. Francesca Fini, "Healing," http://www.francescafini.com/#!video2/cabw/.

27. Patrick Lichty, "La Cura—Salvatore Iaconesi's Glioma," http://www.thingiverse.com/thing:30987/.

28. See, for example, Jane Wakefield, "Crowd-Sourcing a Cure for Cancer Through the Internet," BBC News, October 15, 2012, http://www.bbc.co.uk/news/technology-19899469/; or Alison George, "Crowdsourcing a Cure for my Brain Cancer," *New Scientist*, October 31, 2012, http://www.newscientist.com/article/mg21628880.300-crowdsourcing-a-cure-for-my-brain-cancer.html/, in which a full hour of phone conversation has been cut to the few and sparse answers that deal with the crowdsourcing process, systematically discarding all other issues.

29. Arthur Frank, *The Wounded Storyteller: Body, Illness and Ethics* (Chicago: University of Chicago Press, 1995).

30. See, among others, Jacqueline Bender, Maria-Carolina Jimenez-Marroquin, and Alejandro Jadad, "Seeking Support on Facebook: A Content Analysis of Breast Cancer Groups," *Journal of Medical Internet Research* 13, no. 1 (2011); Joe Dumit, "Illnesses You Have to Fight to Get: Facts as Forces in Uncertain, Emergent Illnesses," *Social Science & Medicine* 62, no. 3 (2006); Gunther Eysenbach, "Medicine 2.0: Social Networking, Collaboration, Participation, Apomediation, and Openness," *Journal of Medical Internet Research* 10, no. 3 (2008); and Hugh Stephens, "Social Media and Engaging with Health Providers," in *Rare Diseases in the Age of Health 2.0*, ed. Rajeev Bali et al. (Berlin: Springer-Verlag, 2014).

31. See Massimiano Bucchi and Federico Neresini, "Science and Public Participation," in *Handbook of Science and Technology Studies*, ed. Edward Hackett

et al. (Cambridge, Mass.: MIT Press, 2008); and Christopher Kelty and Aaron Panofsky, "Disentangling Public Participation in Science and Biomedicine," *Genome Medicine* 6, no. 1 (2014).

32. Steven Epstein, "Patient Groups and Health Movements," in *Handbook of Science and Technology Studies,* ed. Edward Hackett et al., 500.

33. See, for example, Carlos Novas, "Genetic Advocacy Groups, Science and Biovalue: Creating Political Economies of Hope," in *New Genetics, New Identities: Genetics and Society,* ed. Paul Atkinson et al. (London: Routledge, 2007).

34. Kate O'Riordan, "Biodigital Publics: Personal Genomes as Digital Media Artefacts," *Science as Culture* 22, no. 4 (2013); Marina Levina, "Googling Your Genes: Personal Genomics and the Discourse of Citizen Bioscience in the Network Age," *Journal of Science Communication* 9, no. 1 (2010); see also Anna Harris et al., "The Gift of Spit (and the Obligation to Return It): How Consumers of Online Genetic Testing Services Participate in Research," *Information, Communication & Society* 16, no. 2 (2013).

35. Murphy, *Seizing the Means of Reproduction.*

36. Victor W. Turner, *The Drums of Affliction: A Study of Religious Process among the Ndembu of Zambia* (Oxford: Clarendon Press, 1968); and "Symbols in African Ritual," in *Symbolic Anthropology: A Reader in the Study of Symbols and Meanings,* ed. Janet L. Dolgin, David S. Kemnitzer, and David M. Schneider (New York: Columbia University Press, 1977).

37. Claude Lévi-Strauss, "The Sorcerer and His Magic," *Structural Anthropology* 1 (1963).

38. Turner, *Drums of Affliction;* "Symbols in African Rituals."

39. Gabriella Coleman, *Coding Freedom* and "The Hacker Conference: A Ritual Condensation and Celebration of a Lifeworld," *Anthropological Quarterly* 83, no. 1 (2010). See also Levy, *Hackers.*

40. Franco Basaglia, "Psychiatry Inside Out: Selected Writings of Franco Basaglia," ed. Nancy Scheper-Hughes and Anne M. Lovell (New York: Columbia University Press, 1987).

41. Murphy, *Seizing the Means of Reproduction.*

42. Nancy Scheper-Hughes and Margaret M. Lock, "The Mindful Body: A Prolegomenon to Future Work in Medical Anthropology," *Medical Anthropology Quarterly* 1, no. 1 (1987).

43. Lévi-Strauss, "The Sorcerer and His Magic."

44. See also the dimensions of participation in Christopher Kelty and Aaron Panofsky, "Disentangling Public Participation in Science and Biomedicine."

45. Novas, "Genetic Advocacy Groups, Science and Biovalue."

46. Tarleton Gillespie, "The Politics of 'Platforms,'" *New Media & Society* 12, no. 3 (2010).

47. David Hess, "Technology- and Product-Oriented Movements: Approximating Social Movement Studies and Science and Technology Studies," *Science, Technology & Human Values* 30, no. 4 (2005); Johan Söderberg and Alessandro Delfanti, "Hacking Hacked! The Life Cycles of Digital Innovation," *Science, Technology and Human Values* 40, no. 5 (2015).

48. Stalder, *Digital Solidarity*.

49. Nico Carpentier, "The Concept of Participation: If They Have Access and Interact, Do They Really Participate?" *CM—Communication Management Quarterly* 6, no. 21 (2011).

50. Nathalie Casemajor et al., "Non-Participation in Digital Media: Toward a Framework of Mediated Political Action," *Media, Culture & Society* 37, no. 6 (2015).

CHAPTER 8

Internet-Mediated Mutual Cooperation Practices

The Sharing of Material and Immaterial Resources

Bart Cammaerts

MUTUAL COOPERATION IS DEFINED HERE as a form of social and/or economic cooperation between citizens based on the sharing of material goods, as well as immaterial resources such as services, skills, or knowledge. They tend to be characterized not only by forms of solidarity, but also by expectations of reciprocity. Besides this, a sense of fairness and social justice tends to be embedded in mutual cooperation practices. Forms of mutual cooperation have always been very common and could be considered participatory, since they inherently involve a social aspect and bring people together through the sharing of resources. As such, mutual cooperation can build communities and foster social cohesion. Recently, it has often been argued that such cooperative practices are on the rise as a result of the 2008 financial crisis and the ensuing austerity policies enacted by many Western democracies.[1] From this perspective, mutual cooperation is often positioned as an important alternative to capitalist exchange models.

Modes or practices of mutual cooperation are—like so many other aspects of social life today—increasingly mediated through the Internet and networked technologies such as computers, tablets, and smartphones. These technologies not only facilitate modes of mutual cooperation, but have increasingly become constitutive of them as well. This implicates the role of networks and questions the common presumption that modes of mutual cooperation need strong ties and social exchange. While this is certainly still the case for certain forms of mutual cooperation, computer- or mobile telephone–mediated networked power can also rely on the aggregate strength of weak, latent, or even nonexistent ties all acting in conjunction and with a predefined purpose.

This chapter presents a typology of different forms of Internet-mediated mutual cooperation practices prevalent today. In doing so, a practice-based approach is foregrounded, emphasizing what various actors do with media and communication tools in relation to sharing.[2] The first distinction that is introduced relates to the nature of what is being shared in the act of cooperation: material goods or immaterial resources. Within these two categories we can further distinguish between initiatives that are: profit-making versus not-for-profit; catering to individualistic motives versus collective ones; predominately top-down versus predominately bottom-up or grassroots; weak tie–reliant versus strong tie–reliant; and Internet-supported as opposed to Internet-based. Before delving into these distinctions, however, the concepts of mutual aid and cooperation will be theorized in more depth.

Theorizing Mutual Aid and Cooperation

There exists a common perception that the development and advancement of humankind is intrinsically linked to competition, individualism, and, ultimately, to the Darwinist dictum of the survival of the fittest. In his 1888 essay "The Struggle for Existence in Human Society," Thomas Henry Huxley—grandfather of Aldous Huxley—argued that a Hobbesian war of "each against all" and the principle of the survival of the fittest created the need for society and structures capable of taming the beast inside us.[3] However, the Russian scientist Pyotr Alexeyevich Kropotkin—well versed in biology and informed by his extensive travels—found it difficult to reconcile himself with such a pessimistic and deterministic view of nature and humanity. Kropotkin's own scientific research into evolution and human nature showed, on the contrary, that:

Usages and customs created by mankind for the sake of mutual aid, mutual defense, and peace in general, were precisely elaborated by the "nameless multitude." And it was these customs that enabled man to survive in his struggle for existence in the midst of extremely hard natural conditions.[4]

Countering the common views that humans are selfish and egoistical, and liable to thrive on exploitation and greed, Kropotkin thus argued that modes of mutual aid and cooperation among equals were much more prevalent in (human) nature than models based on competition. Humans, Kropotkin pointed out, tend to develop a sense of justice and create social communities because of a need for sociability.[5] As paleontologist Stephen Jay Gould explains it, Kropotkin "did not deny the competitive form of struggle, but he argued that the cooperative style had been underemphasized and must balance or even predominate over competition in considering nature as a whole."[6] Or to put it in Kropotkin's own words: "Mutual aid is as much a law of nature as mutual struggle."[7]

However, mutual aid and cooperation do not follow only from naïve altruism. These social customs come about to ensure that individuals do not free-ride, and to protect the prosocial aspects of the collective from the antisocial behavior found in some of its parts. While unselfishly motivated altruism is seen to play an important part in modes of cooperation based on mutual aid, cooperation can also be seen as an investment strategy, motivated by a belief in reciprocity. While there are forms of sharing and cooperation that do not necessarily invoke expectations of reciprocity, often a kind of "tit-for-tat" expectation can be observed, rewarding acts of cooperation and punishing failures to reciprocate, ultimately ensuring that everybody in the collective "prospers from mutual co-operation."[8]

This ties in with the work of anthropologist Marcel Mauss who argued that traditional gift economies are based on reciprocity.[9] Mauss concluded that small-scale gift economies tend to reinforce the position of those who are able to give more compared to others, ultimately leading to the cementing of social dependencies and the reinforcement of existing power relations. Despite this, Sennett argued that cooperation is "an exchange in which the participants benefit from the encounter." Furthermore, he points out that cooperation is "instantly recognizable, because mutual support is built into the genes of all social animals; they cooperate to accomplish what they can't do alone."[10]

As examples of human modes of mutual aid and cooperation, Kropotkin mentions "workers unions"; "strikes"; "co-operative societies"; "associations based on the readiness to sacrifice time, health, and life if required"; and "the countless societies, clubs, and alliances, for the enjoyment of life, for study and research, for education."[11] He goes on to state that these examples demonstrate that mutual cooperation "represents an immense amount of voluntary, unambitious, and unpaid or underpaid work" by a considerable part of the population.[12]

These arguments challenge common perceptions that development and advancement are intrinsically linked to competition, individualism, and ultimately the survival of the fittest. Continuing to survey this perspective, we could also refer to the work of Elinor Ostrom, who examines how people come together to solve "common pool resource problems."[13] She challenges the traditionalist game theory perspective that frames resource management problems as a series of games with clear and identifiable winners and losers. In reality, Ostrom argues, groups of people in many situations around the world negotiate cooperative approaches to manage resources without there necessarily being winners and losers.

Mutual aid and cooperation tends to operate in a space in between the market economy and the state-subsidized sector and is therefore seen to be mostly situated in the realm of civil society.[14] However, this does not mean that there are no connections between mutual cooperation initiatives and market actors or state institutions. Indeed, as Sennett argues, the "combination of cooperation and competition appears in economic markets, in electoral politics and in diplomatic negotiations."[15]

The notion of mutual aid and cooperation—as well as an engagement with Kropotkin's work and ideas—can be found in diverse and wide-ranging literatures relating to development in the global South, self-help in terms of health issues, (natural) disaster relief efforts, and the establishment of cooperatives and activist autonomous zones. In this chapter, however, we will address first and foremost mutual cooperation through sharing and giving. Another focal point will be the role of the Internet in facilitating and constituting mutual cooperation practices.

Mutual Cooperation and Networked Technologies

As in so many other domains of the social and the cultural, networked technologies such as the Internet and, increasingly, mobile technologies like smartphones play a pivotal role in terms of facilitating and increasingly constituting mutual aid and cooperation. Networks are often understood as composed of direct, indirect, and latent connections between individuals and/or organizations in collaborative endeavors, and they are increasingly mediated in a variety of ways.[16] In network theory, the precise nature of these connections or ties between different nodes is qualified in terms of strength. In relation to the Internet, many scholars refer to its ability to mobilize and connect people with weak ties, but they also identify its ability to engender new tensions between the social/collective and the individual.[17]

From this perspective, the strength of weak ties lies in the ability of individuals and organizations to draw support from their weak tie networks, which offer a range of experiences, information, and resources.[18] The strength of strong ties, conversely, lies in more determined motivations and higher degrees of loyalty. Besides weak and strong ties, Caroline Haythornthwaite identifies a third type: networked technologies, she argues, "support latent social network ties, used here to indicate ties that are technically possible but not yet activated socially."[19]

The weak and latent ties enabled by the Internet facilitate citizen participation and engagement in mutual cooperation, and also help participants manage their degree of involvement. Because the Internet can also be seen as a nonhuman actor that is not only a mediator, but also an intermediary,[20] modes of mutual cooperation devoid of social interaction and exchange can also emerge.[21] The Internet can thus be seen to both facilitate traditional modes of mutual cooperation and produce new forms of mutual cooperation that are Internet-embedded.

This can be mapped onto the distinction Jeroen Van Laer and Peter Van Aelst have introduced between *Internet-supported* and *Internet-based* social movements.[22] Regarding the former, we could refer to many traditional mutual cooperation practices that are Internet-mediated—for example, text-based and Internet-supported self-help groups or support networks, and networked alliances between cooperatives. Along the same lines, many emergent forms of sharing material goods—distinguishable from traditional gift relationships—are currently mediated through the Internet.[23]

In the early days of the Internet, Peter Kollock pointed out that the emergence of digital technologies and the virtual communities they enabled served to considerably reduce the costs associated with giving and sharing.[24] This can, among others, be explained by a process which Ernest Mandel called "total automation"—that is, infinite reproduction without additional labor costs.[25] Temporal and special constraints and boundaries can also be overcome potentially. Furthermore, many forms of Internet-based sharing—and some forms of Internet-supported sharing—are embedded into an ethos that supports the commons and the public domain.[26]

At the same time, Nicholas A. John quite rightly points to the coexistence of commercial and noncommercial forms of sharing.[27] On the one hand, we can observe modes of collaborative consumption whereby the shared goods are owned by a third party and tend to be profit-making—what we might call share.com. On the other hand, modes of mutual sharing emerge whereby the shared goods are owned by those participating in the collaborative consumption initiative, or whereby the sharing of private goods serves a public good; such initiatives can be seen as not-for-profit forms of sharing/lending—what we might call share.org. Apart from enabling these forms of material goods–focused sharing, networked technologies such as the Internet and mobile devices also facilitate and support the communicative sharing of immaterial resources such as support, knowledge, and digitalized cultural products.[28]

Regarding digitized content, it should be acknowledged that there is a distinct materiality to immaterial resources such as digitalized music, films, or software. However, in this chapter I consider it under the heading of immaterial resources precisely because digitalization enables these "products" to be easily and freely distributed and shared on a global scale, disrupting the capitalist logic of exchange value and commodification so reminiscent of material cultural products. As Mark Poster argued:

> When cultural objects are digitized, they take on certain characteristics of spoken language. Like an oral sentence or a song, digitized voice is easily and with little cost reproduced by the networked computer user. We do not say of someone who repeats a sentence out loud, that he or she is a consumer of that sentence. The model of consumption does not fit practices of speech or singing.[29]

Toward a Typology of Internet-Mediated Mutual Cooperation Practices

One of the first major distinctions that can be made regarding Internet-mediated mutual cooperation practices relates to the nature of what is being shared. Some of these initiatives are geared toward the sharing of material resources, while others share immaterial resources or services that are nonphysical.

In the first category, a variety of material resources are being shared: physical goods, energy, housing, and financial resources for investment—that is, capital. Some of these are related to consumption, others to production. The second category—the sharing of immaterial resources—involves the sharing of expertise, knowledge, digital content, and software code.

Internet-Mediated Mutual Cooperation through the Sharing of Material Resources

Many of the Internet-mediated mutual cooperation initiatives are geared toward the sharing of material resources, thereby connecting the virtual to the material in a very concrete way. Many of these initiatives are focused on sharing practices in terms of consumption, but some are also geared toward stimulating production.

Consumption

There is a plethora of examples of Internet-mediated platforms that facilitate mutual cooperation practices related to consumption. Some refer to these forms of mutual cooperation as "collaborative consumption" or, more broadly, "the collaborative economy."[30] Without aiming to be exhaustive, I will address the following practices here: the sharing of expensive goods, the sharing and gifting of cheap goods, group buying, and the sharing of space.

When addressing the sharing of expensive goods, consider that more and more people tend to not own a car, but instead use car-sharing services.[31] For example, Zipcar markets itself as an "alternative to traditional car rental and car ownership."[32] Conditions vary from one country to the other, but this is clearly a for-profit initiative. Zipcar boasts 790,000 members, owns over 10,000 vehicles, is active in more than twenty cities, and maintains a presence on more than 300 college campuses. In 2012, it reported revenues of $279 million, after which Avis, a giant in the car hire business, bought up Zipcar in March 2013. There are also, however, car-sharing services being offered on a more not-for-profit basis or by the state through their provision of public transport. An example of the latter is the Belgian public rail company NMBS/SNCB's car-sharing service Cambio. An excellent example of the former is the nonprofit organization City

CarShare, which is based in San Francisco and provides hybrid and electric cars to its members.

Besides cars, other expensive goods are also increasingly shared. These services are, however, not targeted at "ordinary" consumers but at the very rich. Through what is called fractional ownership, affluent people can co-own a private jet or a luxury yacht. For example, Flexjet, a subsidiary of Bombardier, offers a "FlexShare" fractional ownership program enabling their customers to co-own a private jet, including access to a pilot.[33] Along the same lines, in return for a sizable amount of money (namely between $80,000 and $164,000 a year), SeaNet Fractional Yachts offers its clientele "a piece of the yachting lifestyle."[34]

Besides large and expensive goods, small and inexpensive goods are also increasingly shared or gifted. In the United States, the website NeighborGoods facilitates the borrowing of relatively cheap but useful goods, from stepladders to bikes, chairs, and DVDs. Some of these platforms are more focused on one particular type of good—for example, on the exchange or lending of books (e.g., BookLending.com). Other initiatives, such as the Freecycle Network, focus on the recycling of goods by connecting people who want to throw things away with those that are looking for free goods. Freecycle purports to keep "good stuff out of landfills" and they have almost eight million members worldwide.[35] However, such not-for-profit initiatives are rather fragile and at times have difficulty sustaining themselves. For example, the website http://www.ecosharing.net (referred to in John)[36] currently exists as an informational blog, but it once hosted a website that aimed to facilitate the exchange, sharing, or even gifting of all sorts of goods at a very local level.

Another distinct category of mutual cooperation with relevance to consumption is the phenomenon of group buying. For example, the north Belgian socialist party (SP.a) and trade union (ABBV) have set up http://samensterker.be (Stronger Together). This website enables consumers and vulnerable groups in local communities to reduce the cost of buying energy products, such as wood, electricity, and fuel oil for heating houses. Through group buying, a much sharper price can be negotiated with energy and fuel companies, considerably reducing the costs for those who participate.[37] However, we can also identify quite a number of for-profit group buying initiatives that utilize this sort of "cooperation"; Groupon comes to mind as undoubtedly the most popular one.

Finally, when it comes to consumption we also have to mention the sharing of space and the phenomenon of couch surfing.[38] The site https://www.couchsurfing.com has over six million members in 100,000 cities, enabling members to get access to cheap accommodations when traveling. Use of the website or its associated apps is free and participants cannot charge for the use of their couch, so to speak. Another example, more top-down and for-profit, is the website https://www.airbnb.com. This company facilitates finding cheap places to stay in over 34,000 cities and 190 countries. It enables people to rent out their unused spaces to earn extra cash. Airbnb takes a commission from both those who rent out and those who pay to stay.

Production

In relation to production, I will focus on mutual cooperation regarding the sharing of capital and provision of credit that supports production. The other side of this coin is debt, and in this context we also find initiatives that share the burden of, or alleviate, debt.

Financial resources are increasingly being shared in view of stimulating production. A recent example of this is the explosion of crowdsourced funding initiatives such as U.S.-based Kickstarter and Wefund, and the UK-based website http://www.crowdfunder.co.uk. These online platforms enable artists, and also NGOs and community organizations, to pitch a project and solicit capital from the general public to fund it, be it a documentary, a performance art piece, the starting of a co-op, or skills training for the unemployed. However, a common critique is that this leads to a kind of populist market model in terms of cultural funding (i.e., only those projects that receive popular support get realized), especially as this phenomenon is often pitched against dwindling or slashing of public/state support for cultural production. Furthermore, many of these websites are designed to generate profit—they are, after all, dot-coms.

The story of http://www.crowdfunder.co.uk, for example, is telling in this regard. On the one hand, the website exhumes a nonprofit community ethos, as this quote exemplifies:

> We believe that community is a vital part of our everyday lives which is why we work with project owners to make sure their ideas benefit those around them rather than themselves.[39]

On the other hand, the site is owned by Keo Digital, which also works for high-profile clients such as the Union of European Football Associations (UEFA), the UK National Lottery, Channel 4, British Gas, and Endemol (the production company behind *Big Brother*). Furthermore, in the small print of the term and conditions, it is mentioned that the platform charges a 5 percent-plus VAT fee on the total of all pledges received from backers of a particular project (if it has reached its target). When considering reciprocity, some projects also promise rewards for backers. For example, if you fund the studio time of a band recording a new CD, you might receive the CD "for free."

We can also identify credit-providing initiatives that are much more blatantly commercial, and even exploitative. We can refer here to peer-to-peer lending sites that enable individuals to lend money to others through a mediating online platform. The UK site http://www.piggy-bank.co.uk describes peer-to-peer lending as "a smarter, fairer and more human way of managing money. It's like borrowing and lending with your friends and family—except there are thousands of people you can lend and borrow with."[40] However, the interest rates are considerably higher than those charged by ordinary banks. PiggyBank pays 12 percent on a yearly basis to those who pay into the scheme, while it charges a staggering 1,355 percent annual percentage rate to those borrowing from the website.

As pointed out above, the other side of credit is debt. As such, there are also mutual cooperation initiatives that focus on the sharing of debt rather than on the provision of capital. The website http://rollingjubilee.org, for example, was set up by Strike Debt, an offshoot of Occupy Wall Street. Historically, a jubilee was a religious event wherein debts were canceled. Rolling Jubilee creatively uses the system whereby banks sell off bad debt, at a fraction of its supposed value, to companies that specialize in aggressively pursuing debtors. Rolling Jubilee buys off debt from banks and subsequently abolishes it, rather than attempting to collect it from the debtors. By 2016 it collected about $700,000 in donations and used the money to cancel almost $32 million worth of debt. In one operation, on January 23, 2013, Rolling Jubilee bought up and canceled over $1.1 million worth of medical debts for the price of $20,979; over one thousand debtors saw their medical debts abolished as a result.[41]

Discussion

Taken together, it becomes apparent that the large majority of the mutual cooperation practices discussed above are in some way or another for-profit, using the sharing economy and mutual collaboration to generate added value. However, some actors also cross-subsidize the not-for-profit sharing practices with commercial activities, and some purely not-for-profit initiatives also exist. Furthermore, regardless of whether it is for-profit or not-for-profit, when it comes to consumption, but also in terms of the stimulation of production, there is often a transactional element to these practices—that is, money changes hands. Reciprocity thus occurs mostly through the paying of fees, and the potential to save money functions as the impetus to participate. When it comes to providing capital or credit, reciprocity is a return on investment. This return might be lower in crowdfunding initiatives than it is in relation to peer-to-peer credit provision, but there is still an expectation of a return or reward.

It is important in this regard to take into account the motivations or incentives offered to participants for engaging in these mutual cooperation practices. While some of these practices facilitate the self-interested and self-executed action of the individual parties to the exchange, others are part of a collectively executed action whose aim is shared by all the parties in common. In the section above it is clear that motives relating to the former dominate to the detriment of collective motives. This points to the cooptation of the discourse of cooperation and community; it comes to denote self-interested actions that are often transactional in nature. In reality, mutual cooperation is revealed as more of an alternative model *of* market exchange, rather than an alternative *to* market exchange itself.

At the same time, however, it has to be acknowledged that running online platforms comes at a cost, especially if they start generating large amounts of online traffic; bandwidth needs to be paid for by someone. Also, the provision of administrative support to facilitate these mutual cooperation practices can be expensive, especially if they start to grow. As a result, there are many not-for-profit initiatives that have serious problems in terms of sustainability as well as scalability. This might also explain why most of the examples highlighted above are organized in a top-down manner, be it by companies, state institutions, or civil society actors such as unions, rather than being truly bottom-up and grassroots.

When addressing the way networked technologies are relevant to mutual cooperation practices involving the sharing of material resources, it becomes apparent that most of these initiatives are Internet-supported—that is, the Internet plays a facilitative role rather than a constitutive one, as is the case with Internet-based initiatives that take place entirely online. This can be explained because of the intrinsic relationship between the virtual and the material—the online and the offline—that is characteristic of the sharing of material resources. Besides this, it is also apparent that the ties between the participants in online platforms that facilitate mutual cooperation in the sharing of material goods and capital tend to be rather weak, latent, or even nonexistent; seldom are the ties strong.

Internet-Mediated Mutual Cooperation through the Sharing of Immaterial Resources

When considering Internet-mediated mutual cooperation practices focused on the sharing of immaterial resources, it becomes apparent that quite a few differences occur, not only at the level of what is being shared but also at the levels of reciprocity, motives for sharing, the way sharing is organized, and the precise role played by the Internet as a mediating platform. In terms of immaterial sharing, we can distinguish three core resources that tend to be shared: (1) support and expertise, (2) knowledge, and (3) digital content and software code. These practices tend to be more thoroughly Web-based than those relating to material distribution; the Internet as a platform is thus more constitutive of the practices, rather than merely facilitative.

Support and Expertise

The online sharing of emotional support is very reminiscent of one of the early examples of mutual aid, namely self-help.[42] A multitude of support groups can be found online. They form virtual communities and deal with all sorts of medical conditions—both physical and mental—but they also provide mutual support to cope with bereavement, relationship problems, rebellious children, debts, addictions, or alternative sexual identities. Initially, many of these support groups formed on Usenet; examples include: alt.support.depression, alt.support.alzheimers, and alt.support. chronic-pain. Today, many of these sorts of support groups use online forums instead. There are also quite a few online groups that share not merely support but also expertise, specific skills, and recommendations. In this regard, we could refer to the many tutorials available online that people increasingly turn to whenever they want to know how something is done. Likewise, the energy that users, consumers, readers, and so on put into the writing of reviews of a wide diversity of (cultural) products can be seen in this light.

While most of these online emotional support and skills-sharing networks are not-for-profit—focusing on linking up peers with similar conditions, issues, or problems—some are also run by professionals who clearly have other aims beyond sharing and cooperation. For example, the website http://www.dailystrength.org, boasting five hundred online support groups and forums focusing on a variety of "life challenges, medical conditions, and mental health issues,"[43] is hosted by Sharecare, a platform that "is dedicated to connecting consumers with industry experts and other knowledgeable consumers in order to educate, empower, and continue the conversations of health."[44] In essence, websites like this function as patient recruiting devices for (health) professionals, which is why experts pay a hefty monthly subscription to be registered on such websites. This is especially apparent when analyzing Sharecare's discourse targeted at health professionals, which promotes the website as a way "to unlock the tools to successfully market your practice and acquire and retain valuable patient relationships."[45]

Similarly, when it comes to people sharing recommendations for consumer goods, restaurants, holiday destinations, books, and so on, these practices have also been appropriated by large corporations in their efforts to pull consumers toward their websites. Think of how reviews function on Amazon.com or the commercial website Goodreads.com, and also the increasingly dominant role played by reviews of restaurants and hotels on websites like Tripadvisor.com.

There are, however, also ways in which time, expertise, and skills are traded in a not-for-profit and more collective manner. In this regard, we can refer to the so-called Community Exchange System (CES), whereby citizens exchange their time, skills, and expertise as peers. Instead of relying on traditional currencies, CES users "exchange and share what [they] have to offer for what others provide using a variety of exchange mechanisms: record keeping, time exchange, direct exchange, barter, swapping, gifting and sharing."[46] The CES provides an alternative currency, mediated by an online platform, for (local) communities—in effect eliminating the

need for traditional money as a mode of exchange. The website https://
www.community-exchange.org hosts about 560 CES groups worldwide.
They are especially popular in Spain (162 groups), the United States
(87 groups), South Africa (39 groups), Finland (41 groups), Australia
(32 groups), New Zealand (30 groups), and Canada (18 groups).[47]

Knowledge

Somewhat related to the above, but also distinct enough, are several initiatives that facilitate the online sharing of knowledge. One very obvious
example that comes to mind here is Wikipedia, the famous online ency-
clopedia which combines top-down with bottom-up knowledge sharing
and lay with expert participation.[48] Despite the open architecture of Wikipedia, over the years it has developed a whole set of rules and guidelines regarding the sharing of knowledge resources. Wikipedia boasts 30 million articles
in 287 languages and 18 billion page views every month.[49] It is noteworthy, in
this regard, that Wikipedia initially started out not as the community-based
project it purports to be today, but as a for-profit venture.[50]

Another example of shared knowledge production relates to the provision of open access as applied to academic publication and knowledge
production. At the time of writing, the Directory of Open Access Journals
(DOAJ) lists about 11,300 journals based in more than 135 countries
across the world.[51] There is also increased pressure from funding bodies —
private as well as public — to make the outcomes of the academic research
they fund accessible to the wider public, instead of just appearing in journals owned by commercial publishers that place content behind paywalls
and beyond the reach of most people.[52]

This has, however, given rise to a great diversity of publishing models
in the realm of open access. For example, gold access models are deemed
to be the most open; they are defined by the Budapest Open Access Initiative as:

> The world-wide electronic distribution of the peer-reviewed
> journal literature and completely free and unrestricted access to
> it by all scientists, scholars, teachers, students, and other curious
> minds. . . . While the peer-reviewed journal literature should be
> accessible online without cost to readers, it is not costless to
> produce.[53]

The sting is in the last part of this quotation. While many open access initiatives work along the community model whereby costs are borne by volunteer labor and creative commons licenses protect the interests of the authors, commercial publishers have in the meantime responded to pressure from research funders by developing their own gold access models. Many of these models transfer the costs of production to the authors by charging them so-called article processing costs (APC).[54] Beyond the gold standard, there are a number of other models, ranging from delayed open access to university repositories that host pre-review versions of articles (cf. http://eprints.lse.ac.uk).

Commercial publishers have, in other words, managed to coopt the open access model in order to generate an additional revenue stream. For example, Cogent Open Access, a subsidiary of Taylor and Francis, recently advertised an International Open Access Week (October 20–26, 2014).[55] Inevitably, this shift has also led to the emergence of dubious journals that aim to take advantage of gold models (see, for example, http://scholarlyoa .com/).

Digital Content and Software Code

The final immaterial resource that is often shared today relates to the digital space itself. It must be acknowledged that more than in the examples referred to above, there is a materiality to the immateriality of digitalized "products." However, when it comes to (1) the sheer extent of digitalized content that is shared and distributed online, and (2) the way in which software code is being shared and/or gifted, it is the nonphysical attributes that make a difference in terms of the opportunities to share widely. These attributes thereby disregard or indeed disrupt (to some degree!) the capitalist value exchange model. While the sharing of digital content has more to do with consumption, at the level of software code the focus is more on production.

As I argue elsewhere, the sharing of culture is not a new phenomenon, but digitalization certainly has increased the scale, speed, and magnitude at which immaterial cultural commodities are shared outside the value chain.[56] While digital file sharing in itself is used in a variety of legal ways, the sharing of copyright-protected content is obviously highly controversial, as it rejects the hegemonic property regime and disrupts the commodification of cultural production. Besides this, and relevant for the discussion in this chapter, peer-to-peer practices that distribute content freely have a kind of built-in reciprocity: While you download material from some sites, you must also upload to others. At the same time, the sharing of digital content and software also occurs through file-sharing websites that are infested with online advertising and provide premium services; in a sense they are "creating added value through the disruption."[57]

The free and open source software (FOSS) movement, which shares and gifts software code, operates very much within the copyright paradigm, just as open source journals do. In this case, copyright law is used not to protect property but to protect the public domain and to ensure that gifted code is not commodified. In this regard, a difference can be observed between so-called copyleft licenses and open source licenses. The latter enables more flexibility for companies to exploit shared software code by offering auxiliary services or by embedding it within propriety code.[58]

Discussion

Practices related to the sharing of immaterial resources seem to be less concerned with profit making than those related to the sharing of material resources. It has to be acknowledged, though, that the sharing of immaterial resources is not entirely devoid of commodification either. Some professionals provide advice and offer their expertise online as a strategy to recruit clients. The publishing industry offers open access only partially or at a price, and websites facilitating file sharing often offer premium services or bombard us with aggressive online advertising. Furthermore, major players in the software industry invest in open source software production because it suits their commercial interests. Despite all this, reciprocity is more characterized by the logic of "share-alike" rather than through fees or transactions.

Likewise, it seems that at the level of the actual motivations to collaborate or cooperate, practices that genuinely serve collective aims and the common good are less prevalent than individualistic reasons geared toward consumption. Not-for-profit online support networks do at times build strong communities that serve collective goals, many open source journals are genuinely concerned with free access to academic knowledge for all, and peer-to-peer software has an inherent collective edge to it. However, the coin also has another side, where individualistic motives reemerge—for example, professionals who provide advice as part of their marketing efforts and commercial publishers who coopt and transform the open access model or download content without reciprocating or sharing alike.

This dual tendency to mobilize collective aims and to subsequently coopt them for individualistic or particularistic ends can also be observed when it comes to the organizational structures that underpin the practices of sharing immaterial resources. It seems that it is much easier to organize such practices in a bottom-up fashion, but at the same time there are clearly also top-down organized initiatives coopting the discourse of bottom-up cooperation.

Mutual cooperation practices that facilitate the sharing of immaterial resources are more Internet-based than Internet-supported. However, the ties between those sharing immaterial resources can potentially be quite strong, especially when we refer to not-for-profit support networks, CESs, or not-for-profit open access journals. This might seem paradoxical, but given that the sharing of immaterial goods is more often organized in a bottom-up, horizontal manner that produces stronger links, this is not entirely unsurprising either.

Conclusions

A number of observations and paradoxes emerge out of this analysis, some of which have already been outlined in the discussion above. When considering the nature of reciprocity, we can see that when material goods are exchanged, reciprocity is often characterized by money changing hands, making it very suitable for commercial exploitation. Cooptation is also prevalent in practices relating to capital and production. Some practices, however, manage to resist this temptation—the sharing or gifting of inexpensive goods, for example. But such forms of small-scale, local sharing practices are highly vulnerable in terms of sustainability. Mutual cooperation practices related to the sharing of immaterial resources tend to operate more on a share-alike basis, and at times the question of reciprocity is even irrelevant.

When assessing motives, a distinction needs to be made between those practices that serve individualistic aims and those that serve the collective aims of all those collaborating. It seems that the former are often more present than the latter—certainly when it comes to sharing material, but also in some cases of the sharing of immaterial resources.

Along the same line, when addressing organizational forms, the sharing of material goods seems most often organized in a top-down manner, which is detrimental to the building of strong ties. With the sharing of immaterial resources, a bottom-up way of organizing is easier to achieve, and this more often, but not always, leads to the establishment of stronger ties. If we consider the role of networked technologies in all this, we can see that they play a more facilitative, supportive role in terms of the sharing of material resources, whereas with the sharing of immaterial resources via Web-based initiatives they are more constitutive; CES groups, which retain links to offline communities, are the exception to this.

When returning to the theoretical foundations of mutual aid and cooperation, and their embeddedness in civil society, we can observe a wide variety of actors with very different intentions active in this realm. If we assess the aptness of mutual cooperation practices as an alternative model capable of escaping from capitalism, it has to be acknowledged that this is only partially possible. Commercial "social enterprise" and top-down models crowd out alternative not-for-profit grassroots models, especially when it concerns the sharing of material and capital resources. Many sharing practices, such as group-buying energy or group-renting cars, are also geared toward "getting a better deal" instead of challenging capitalism per se. As such, many of the sharing and collaborative practices discussed in this chapter do not mobilize collective goals, but rather individualistic ones.

When we assess the anarchist/autonomous antecedents of mutual aid and cooperation, we also observe that structures such as states, large NGOs, trade unions, political parties, and indeed also companies tend to organize or enable most of these sharing practices, and that a lack of such structures tends to impede the further development—and above all the scalability—of such initiatives.

At the same time, and certainly in the context of austerity politics, we can also observe how anarchism and neoliberalism meet in some of these mutual cooperation practices. The discourse inherent to such practices is one of self-organization, of citizens taking responsibility to care for one another, to deliver social and cultural services that used to be provided by the (welfare) state. Also at the level of motives, we can observe that often notions such as the "collaborative economy," "mutuality," and "sharing is caring" are being appropriated by individuals and (social) enterprises for personal gain or to generate added value, thereby reducing mutual cooperation to alternative forms of market relations, plus a bit of charity.

Finally, all of this points to the posthegemonic stage we currently find ourselves in, wherein it has become almost impossible to escape capitalism and neoliberalism.[59] Even the bottom-up and more resistant modes of sharing, such as Rolling Jubilee or the creative commons and free software movements, all operate squarely within capitalism and its rules of engagement, using capitalism against capitalism, gaming the system from within the system rather than radically challenging it or constituting a radical alternative to it. Furthermore, the motive for sharing digital content is often—but not always—not having to pay for it.

It is important, however, not to succumb here to what Gramsci denoted as the pessimism of the intellect.[60] It remains important to acknowledge and reiterate the real potential for alternative models of exchange and the sharing of common resources, be they material or immaterial. As such, we require further research in terms of the sustainability of genuine not-for-profit initiatives, their scalability, and the collective benefits to its potential participants.

Notes

I would like to thank the participants in a (failed) bid I led for an EU research project on mutual cooperation. I am also grateful to the reviewers as well as the editors of this book for their very useful and constructive comments and suggestions.

1. Manuel Castells and Bregtje van der Haak, *Aftermath: Life beyond the Crisis* (DVD-ROM) (Hilversum: VPRO, 2011).

2. Nick Couldry, "Theorising Media as Practice," *Social Semiotics* 14 (2004): 115–32.

3. Thomas Henry Huxley, *Evolution and Ethics and Other Essays* (New York: D. Appleton & Company, 1899).

4. Pyotr Alexeyevich Kropotkin, *Evolution and Environment* (London: Black Rose, 1995 [1903]), 48–49.

5. Ibid., 68–69.

6. Stephen Jay Gould, *Bully for Brontosaurus* (London: Penguin Press, 1991), 335–36.

7. Pyotr Alexeyevich Kropotkin, *Mutual Aid: A Factor of Evolution* (London: Freedom Press, 2009 [1902]).

8. For a discussion of forms of cooperation that do not depend upon expectations of reciprocity, see Russell Belk, "Why Not Share Rather Than Own?" *Annals of the American Academy of Political and Social Science* 611 (2007): 126–40. For discussion of tit-for-tat motivation, see Richard Dawkins, *The Selfish Gene* (Oxford: Oxford University Press, 2006), 219.

9. Marcel Mauss, *The Gift: The Form and Reason for Exchange in Archaic Societies* (London: Routledge, 2000 [1950]).

10. Richard Sennett, *Together: Rituals, Pleasures and Politics of Cooperation* (New Haven, Conn.: Yale University Press, 2012), 5.

11. Kropotkin, *Mutual Aid.*

12. Ibid., 232.

13. Elinor Ostrom, *Governing the Commons: The Evolution of Institutions for Collective Action* (Cambridge: Cambridge University Press, 1990), 7.

14. Jean L. Cohen and Andrew Arato, *Civil Society and Political Theory* (Cambridge, Mass.: MIT Press, 1992).

15. Sennett, *Together.*

16. See Tiziana Terranova, *Network Culture: Politics of the Information Age* (London: Pluto Press, 2004) and Christian Fuchs, *Internet and Society: Social Theory in the Information Age* (London: Routledge, 2008).

17. Lee Rainie and Barry Wellman, *Networked: The New Social Operating System* (Boston, Mass.: MIT Press, 2012).

18. Mark Granovetter, "The Strength of Weak Ties: A Network Theory Revisited," in *Social Structure and Network Analysis,* ed. Peter V. Marsden and Nan Lin (Beverly Hills, Calif.: Sage, 1982), 105–35.

19. Caroline Haythornthwaite, "Social Networks and Internet Connectivity Effects," *Information, Communication & Society* 8 (2005): 125–47.

20. See, by way of comparison, Bruno Latour, *Reassembling the Social: An Introduction to Actor-Network-Theory* (Oxford: Oxford University Press, 2005).

21. Andrea Wittel, "Qualities of Sharing and Their Transformations in the Digital Age," *International Review of Information Ethics* 15 (2011): 3–8.

22. Jeroen Van Laer and Peter Van Aelst, "Internet and Social Movements Action Repertoires: Opportunities and Limitations," *Information, Communication & Society* 13 (2010): 1146–71.

23. See, by way of comparison, Belk, "Why Not Share Rather Than Own?"

24. Peter Kollock, "The Economies of Online Cooperation: Gifts and Public Goods in Cyberspace," in *Communities in Cyberspace,* eds. Marc A. Smith and Peter Kollock, (London: Routledge, 1999), 220–42.

25. Ernest Mandel, *Late Capitalism* (London: New Left Books, 1975).

26. See Elinor Ostrom, *Governing;* Richard Stallman, "The GNU Operating System and the Open Source Revolution," in *Open Sources: Voices from the Open Source Revolution,* ed. Chris DiBona, Sam Ockman, and Mark Stone (London:

O'Reilly and Associates, 1999), 53–70; and Lawrence Lessig, *Free Culture: How Big Media Uses Technology and the Law to Lock Down Culture and Control Creativity* (New York: Penguin Press, 2004).

27. Nicholas A. John, "The Social Logics of Sharing," *Communication Review* 16 (2013): 113–31.

28. Bart Cammaerts, "Disruptive Sharing in a Digital Age: Rejecting Neoliberalism?" *Continuum: Journal of Media and Cultural Studies* 25 (2011): 47–62.

29. Mark Poster, "Consumption and Digital Commodities in the Everyday," *Cultural Studies* 18 (2004): 409–23.

30. On the subject of "collaborative consumption," see Rachel Botsman and Roo Rogers, *What's Mine Is Yours: The Rise of Collaborative Consumption* (New York: Harper Business, 2010). Regarding the "collaborative economy," see Vasileios Kostakis and Michel Bauwens, *Network Society and Future Scenarios for a Collaborative Economy* (Basingstoke: Palgrave Macmillan, 2014).

31. See Yochai Benkler, "Sharing Nicely: On Shareable Goods and the Emergence of Sharing as a Modality of Economic Production," *Yale Law Journal* 114 (2004): 273–358, and Susan A. Shaheen and Adam P. Cohen, "Growth in Worldwide Carsharing: An International Comparison," *Transportation Research Record: Journal of the Transportation Research Board* 1992 (2007): 81–89.

32. "Car Sharing, an Alternative to Car Rental and Car Ownership—Zipcar," *Zipcar,* http://www.zipcar.com/.

33. "FlexShare Program Offers Flexjet Owners More Customization Options," *Flexjet,* October 29, 2012, http://www.flexjet.com/news-and-events/press-releases /231/.

34. "Yacht Charter | Yacht Ownership Programs," *SeaNet Fractional Yachts,* http://www.seanetco.com/available-programs/.

35. "The Freecycle Network," http://www.freecycle.org/.

36. See John, "Social Logics," 119.

37. "Bespaar nu met Samenaankopen—Samensterker," http://www.samen sterker.be/.

38. Devan Rosen, Pascale Roy Lafontaine, and Blake Hendrickson, "Couch-Surfing: Belonging and Trust in a Globally Cooperative Online Social Network," *New Media & Society* 13 (2011): 981–98.

39. "A Page about Crowdfunder's Ethos," http://www.crowdfunder.co.uk/about /our-ethos/.

40. "How Short Term Loans and Payday Loans Work at PiggyBank," PiggyBank, http://www.piggy-bank.co.uk/how-piggy-bank-works/. (Note that this quote has since been removed from the referenced page).

41. "Purchase Summary," Rolling Jubilee Fund (January 23, 2013), http:// rollingjubilee.org/assets/docs/debt-buy-summary_02.pdf.

42. Colin C. Williams and Jan Windebank, "Self-Help and Mutual Aid in Deprived Urban Neighbourhoods: Some Lessons from Southampton," *Urban Studies* 37 (2000): 127–47.

43. "Online Support Groups," DailyStrength, http://www.dailystrength.org /support-groups/.

44. "What Is Sharecare? Your Health, Your Body, You," Sharecare, http://www .sharecare.com/question/what-is-sharecare/.

45. "Sharecare Customer Support—Frequently Asked Questions," Sharecare, http://www.sharecare.com/static/faq/.

46. "Welcome to the Community Exchange System," Community Exchange System, https://www.community-exchange.org/.

47. This data was collected at the end of January 2014.

48. Heng-Li Yang and Cheng-Yu Lai, "Understanding Knowledge-Sharing Behaviour in Wikipedia," *Behaviour & Information Technology* 30 (2011): 131–42.

49. "Wikipedia," Wikipedia, last modified January 12, 2015, http://en.wikipedia .org/wiki/Wikipedia/.

50. Dan O'Sullivan, *Wikipedia: A New Community of Practice?* (Farnham: Ashgate, 2009).

51. "Directory of Open Access Journals," DOAJ, http://www.doaj.org/.

52. Craig Brierly, "Wellcome Trust Strengthens Its Open Access Policy," Wellcome Trust (June 28, 2012). http://www.wellcome.ac.uk/News/Media-office /Press-releases/2012/WTVM055745.htm.

53. "Read the Budapest Open Access Initiative," Budapest Open Access Initiative, http://www.budapestopenaccessinitiative.org/read/.

54. Typically, APCs range from $1,500 to $5,000, depending on the discipline and rankings of the journal.

55. "Cogent OA," Taylor & Francis Online, http://cogentoa.tandfonline.com/.

56. Cammaerts, "Disruptive Sharing."

57. Ibid., 58.

58. Johan Söderberg, *Hacking Capitalism: The Free and Open Source Software Movement* (London: Routledge, 2007).

59. Bart Cammaerts, "Neoliberalism and the Post-Hegemonic War of Position: The Dialectic between Invisibility and Visibilities," *European Journal of Communication* 30 (2015): 522–38.

60. Anne Showstack Sassoon, *Gramsci and Contemporary Politics: Beyond Pessimism of the Intellect* (London: Routledge, 2000).

· III ·

Participation under Surveillance

CHAPTER 9

Big Urban Data and Shrinking Civic Space

The Statistical City Meets the Simulated City

Kate Crawford

*Imagine knowing practically every detail about a city. The state of its
infrastructure, its inhabitants, its environment are all known to you,
to high resolutions in time and space. You are able to fuse physical
data streams with socio-economic ones: transport data tells you where
people are going, sales and transaction data tells you what people are
going to see or do or buy, tweets and social networks tell you how they
feel about it—and tell you how those feelings change, minute by
minute, as a few drops of rain fall or the sun comes out. . . . Therein is
the potential of Big Data. A city that worked like that could really get
up and run.*

—Sallie Ann Keller, Steven E. Koonin, and Stephanie Shipp,
"Big Data and City Living—What Can It Do for Us?"

I N MAY 2013, Boston held a two-day outdoor music festival called Boston Calling. It featured the National, Cults, Dirty Projectors, the Shins,
and dozens of other well-known indie and alt-rock bands. It drew thousands of music fans to the forecourt of City Hall. But what the attendees
and even the concert promoters didn't know was that they were unwit-
ting participants in a massive experiment. The city of Boston was beta
testing a live data surveillance and facial recognition system on every
person present. It was described as an "urban laboratory," designed by
and licensed from IBM, a combination of its Smart Surveillance System
(SSS) and Intelligent Video Analytics (IVA) software. A memo from IBM

indicated that they were using "Face Capture" on persons of interest, which was defined as "anyone who walks in the door."[1]

The only reason this large-scale crowd surveillance pilot is public knowledge is because of an investigative piece by three journalists who found documentary evidence while "searching the deep web."[2] They found identifying images of the festival attendees, marked with a score out of one hundred for color of skin, and noting if they wore eyeglasses or were balding. After they contacted the office of Boston Mayor Marty Walsh, press secretary Kate Norton made a statement:

> The purpose of the pilot was to evaluate software that could make it easier for the City to host large, public events, looking at challenges such as permitting basic services, crowd and traffic management, public safety, and citizen engagement through social media and other channels. These were technology demonstrations utilizing pre-existing hardware (cameras) and data storage systems.[3]

The combination of efficiency and public safety has a very particular resonance in this context. Only a month before Boston Calling, the Tsarnaev brothers had carried their backpacks into Back Bay and bombed the finish line of the Boston Marathon. Police had video footage of the brothers but could not identify them for several days. The police failure may have prompted the city to show it could marshal more technological resources to greater effect, and render its citizens legible during mass public events. Given this context, there was increased interest in expanding the already burgeoning infrastructure of surveillance. As Norton explained, it was a combination of existing systems (cameras and data storage) with new capacities (social media mining, facial recognition, and crowd management). If there was a time when the city of Boston was most willing to test all the affordances of crowd surveillance, it was directly after the marathon bombings.

But it also reveals a kind of politics of simulation—it was an en masse facial recognition "pilot test." Why ask people's permission, or even inform the concert promoter, if it is just testing a model? It echoes a moment in DeLillo's *White Noise*, the great sardonic novel of mediation, where the city decides to stage an evacuation as "a chance to use the real event in order to rehearse the simulation":

"What does SIMUVAC mean? Sounds important."

"Short for simulated evacuation. A new state program they're still battling over funds for."

"But this evacuation isn't simulated. It's real."

"We know that. But we thought we could use it as a model."

"A form of practice? Are you saying you saw a chance to use the real event in order to rehearse the simulation?"

"We took it right into the streets."[4]

It's a strange foreshadowing of what happened in Boston. The Boston Calling pilot was a moment where a city used a "real event" in order to rehearse a simulation of crowd detection. Somewhere, the line between real and simulated situations blurs, prompted by a circular logic that always justifies any system of greater data collection, processing, and analysis. Thus city surveillance systems promise greater public safety at a lower cost—be it in regard to terrorism, traffic management, or some unimagined challenge.

This chapter considers the confluence of two logics: the technologies of simulation used to test out scenarios for law enforcement and the military, and big data systems designed to increase urban efficiency. Large-scale urban simulations began in off-site locations, military bases, and custom-made environments, but these experimental projects are now being conducted in the lived environments of millions of people. The justifications for big data cities sit comfortably within the logics of securitization and efficiency, but in many ways they remain outside of traditional conceptions of participatory civic life and accountability.

Participation versus Permacrisis

"Cities are the new laboratories of democracy," said former mayor of New York City Michael Bloomberg in a public address in 2013.[5] "If an innovative program or policy can work in one city, it can spread across the country and even across the globe," he continued. Bloomberg's vision is that cities are testbeds where ideas can "take root locally, and then hopefully spread." This understanding of a city as a lab is now commonplace, and it underlies the many urban data experiments being conducted around the world, including in London, Barcelona, and Amsterdam. While the term "smart city" already brings with it a kind of dark irony (intelligence for whom?), it is also quite indistinct in meaning. It's a designation that is used to point to a wide range of technocratic forms of city management, where the city becomes a "totalizing sensory environment in which human actions and reactions, from eye movements to body movements, can be traced, tracked, and responded to."[6] Mass data gathering is deployed in the context of improving traffic flow, contributing to public health by tracking infectious disease, and developing "predictive policing" methods where crime hotspots are monitored to prevent crime before it happens.[7]

Some custom-built smart city examples, such as New Songdo in Korea and Masdar City in the United Arab Emirates, are new cities being designed from the ground up, and will be studded with sensors and information-processing capacities. But they are exceptions in every sense: Unlike other cities, they have no incumbent human population, existing infrastructure, or ways of being. They are unfinished and uninhabited. We are looking in the wrong place if we consider these as exemplars of the smart city project. So we need to search elsewhere to understand the models and assumptions being tested in hundreds of existing cities—forming a more generalized and dispersed set of techniques that already affect many millions of people around the world. These techniques are part of a shift toward what I'm calling "big urban data": a shorthand that is meant to distinguish these practices from the short list of bespoke smart cities and to draw connections between the logics of big data as they have emerged in the technology sector and those in the built environment. But the act of naming is less important than the ability to recognize the underlying goal, which is to gather and centralize data from a range of personal devices, sensors, and cameras and use this data to analyze, predict, and design interventions on human populations.

Currently there are hundreds of other cities that are testing or already running systems of sensor tracking, cross-platform data analysis, and facial recognition. This vision of the data-managed city is based, as Adam Greenfield observes, on a problematic "unreconstructed logical positivism, which, among other things, implicitly holds that the world is in principle perfectly knowable, its content enumerable and their relations capable of being meaningfully encoded in the state of a technical system, without bias or distortion."[8] Some of these programs of vast data gathering are matters of public record, but the majority are not. As in the case of the Boston Calling festival, programs are found out after the fact, enabling little in the way of a public debate about their suitability or acceptability, either before or afterward. The ideological commitment to data-centered city management has, thus far, been largely unchallenged and unchecked. If cities are meant to be the new laboratories of democracy, there has been very little in the way of a critical assessment of the model of "democracy" that underlies these experiments.

There is now a sense of inevitability that cities are moving toward blanket instrumentation, with barely visible spiderwebs of cameras, sensors, and networks linking across all public (and much private) space. But it is worth considering where this sense of inevitability came from. Partly it's the massive investments in smart cities from technology and infrastructure companies like IBM, Cisco, and Siemens. It's also due to the pressure on cities to be more efficient with their resources—to use data collection and analysis to make their services more targeted. This occurs within a broader political and rhetorical environment where cities are encouraged to run more like businesses; it is suggested they should be operated for maximum "efficiency" and minimal costs—for maximum return on investment. In many cases, this has meant forming public–private partnerships where commercial entities are gathering data on citizens as part of a broader commitment to "innovation" and support for local businesses.

But perhaps more than any other single causal agent in the last decade and a half, a key driver of big urban data has been a powerful sense of permanent crisis in large cities. This "permacrisis" of security was accelerated by the terrorist events of 2001 in New York, 2005 in London, and 2008 in Mumbai. Beyond terrorist events, many cities are increasingly vulnerable to natural disasters, resource shortages, and rapid increases in population. Responses to the permacrisis have often been technological: promises of increased security and legibility through networked awareness. Fundamentally, the big data vision of cities is undergirded by the belief that with more data comes more control. As Adam Greenfield writes, the smart city project is founded in "a discomfort with unpredictability, a positive terror of the unforeseen and emergent—in short, a palpable nervousness about the urban itself. This is why the apparatus of ubiquitous sensing, data-mining and predictive analytics exists in the first place."[9]

The turn to big urban data is deeply connected to the sense of permacrisis, and this discomfort is in turn connected with risk in urban environments. In a state of ongoing crisis, the usual forms of civic participation can be suspended—put on hold until things are returned to a more normal state. Yet a return to "normal" is indefinitely suspended. Further, the apparatus of sensing and data analysis can never "resolve" the permacrisis: In some ways, the exponential increase in data streams can itself be an overwhelming force that requires new forms of management. As was the case with the Boston bombings, important existing information is lost in the ocean of new data sources. But despite this, new sensing systems are designed to be always on, alert to everything. This belief can be found in many of the pitch documents that companies have used to sell the idea of distributed sensing systems to government. One clear example comes from IBM, in its smart city pitches to Washington, D.C., in 2008. The pamphlet speaks of "building an ongoing and systemic listening model."[10] The natural corollary to the permacrisis model is the constant presence of technologies of listening.

Listening Like a State

In some ways, this is just the latest twist in centuries of state-driven schemes of legibility and control. In *Seeing Like a State,* Scott articulates four elements which are combined in the "most tragic episodes of state-initiated social engineering": administrative ordering of nature and society, belief that maximizing technical progress maximizes productivity, state willingness to use coercive power, and a civil society that is unable to resist. In postrevolutionary France, Haussmann's retrofitting of Parisian streets was as much about introducing "military control of these insurrectionary spaces" where civilian resistance had been centered—mainly in the densely packed working-class areas—as it was about improving quality of life in the city. Modern order meant new, wide avenues, which would allow for easy movement between the subservice areas and the barracks. This produced other benefits, including improvements to water supply and public health. But the cost was disproportionately felt by the poor, who were displaced in the hundreds of thousands from the center of the city toward the periphery.[11]

In more recent history, New York in the early twentieth century was the site of considerable planning power and little chance for public resistance. At the height of Robert Moses's power as the chief urban planner of New York, he marshaled enormous financial resources to produce a dramatic reshaping of the city. He built parks and freeways, swimming pools and bridges. But his work also fundamentally reconceptualized the civic space of the city. The large-scale experiments that Moses conducted on New York gave preferential treatment to some communities over others. For example, when the construction of new roads required the demolition of thousands of dwellings, he would select areas that had high African American and immigrant populations. And this was done without having adequate support for resettling the residents.[12] When building public swimming pools, he and his senior advisor discussed ways in which to keep African American families away. These effects—intentional and otherwise—produced a particular vision of democracy in infrastructure. There are ongoing debates about the relationship of the built environment to politics—and the relationship of material form to social content—but for our purposes, we can see that Moses's approach acts first as an example of highly centralized city-making, where community consultation is a low priority and power is concentrated in the hands of the few.[13] But Moses also championed the turn to data: His New York was a prototype of the statistical city.

In her 1961 critique of city planning, Jane Jacobs observes that theorists of the conventional modern city have been unable to identify what kind of "problem" they are addressing with their large-scale city schemes. In some cases, they have mistaken cities as "problems of simplicity and of disorganized complexity," rather than thinking of them as something closer to a system in the life sciences, places of "organized complexity."[14] In the latter view, cities are intensely interconnected and complex, and cannot be broken down into individual monadic elements.

> Planners began to imitate and apply these analyses precisely as if cities were problems in *disorganized* complexity, understandable purely by statistical analysis, predictable by the application of probability mathematics, manageable by conversion into groups of averages [emphasis added].[15]

She described this kind of city planning as a celebration of "statistics and the triumph of the mathematical average."[16] Jacobs diagnosed a problem that has been greatly exacerbated by big data approaches. By emphasizing probability statistics as a way to understand cities, the complex problems inherent in the large-scale relocation of citizens to make way for highways became simplified in ways that were very damaging to already marginalized populations.

In the form of statistics, these citizens were no longer components
of any unit except the family, and could be dealt with intellectually
like grains of sand, or electrons or billiard balls. The larger the
number of uprooted, the more easily they could be planned for on
the basis of mathematical averages."[17]

And here, in essence, is a core article of faith in big urban data programs: that as data sets become larger and more interconnected, a necessarily more "accurate" and "rational" vision will appear. But it can disguise
a narrow form of statistical thinking that prioritizes only those values that
can be quantified. Furthermore, the mythical power of big data is such
that decisions are less likely to be questioned or critiqued when they are
dressed in the majestic raiment of vast data sets and visualizations. This
tactic was already underway in the mid-twentieth century, and was something that Jacobs fiercely criticized:

It became possible also to map out master plans for the statistical
city, and people take these more seriously, for we are all accustomed
to believe that maps and reality are necessarily related, or that if
they are not, we can make them so by altering reality.[18]

She notes that this data-driven worldview could be applied to all manner
of city functions, from city traffic to parks to culture—all "convertible
into problems of simplicity" from an "Olympian" viewpoint.[19] But this
approach produced a fundamental misunderstanding, in her view, of the
complexity of the problems at hand and the diversity of the communities
being treated with a single, data-led approach. In some ways, Jacobs was
one of the first critics of big data in an urban context.

Yet despite the influence of Jacobs's work, the Olympian view lives on.
It can be seen both in the range of justifications used for a shift toward en
masse data gathering in cities and towns, as well as in the growing number
of commercial data platforms that overlay contemporary cities. For example, Uber, the car ridesharing service, has a dashboard view that allows
senior company managers and partners to see all the drivers and passengers moving through a city at one time. At their private events they like to
demonstrate this system, sometimes even displaying the names of the
passengers as they are driven between locations. They call this the "God
View."[20]

At one level big data planning is attempting to see the interconnectedness of certain systems by bringing together data sets from social media, from the census, and from other forms of what are euphemistically called "passively gathered" data sets, like taxi and public transport data. The optimistic hope is that this will help identify inefficiencies that can be addressed and problems that can be solved, and offer insights into how citizens really think and feel. But these are assumptions that often belie the nature of the data sets being gathered and what they represent. It is a repetition of the same problem that Jacobs recognized, and one that multiplies the problem across domains: Social media do not represent "what a city thinks," nor does the combining of data sets bring us a full picture of "ground truth." Each data set has its own form of inbuilt bias, and by bringing sets together we may create an illusion of completeness that masks the multiple gaps in the individual component sets. The God View is tainted by what it *can* and *can't* see, and by the way it further concentrates power in problematic ways.

Algorithmic Listening

If the culture of top-down data experiments means that the public is not able to participate meaningfully in the evolution of cities, or determine the way they are personally recorded and sold on as data, then what does participation in a big data city look like? In 2008, as part of a pitch for smart cities, IBM stressed the need to build "an ongoing and systemic listening model."[21] In other words, in addition to modeling cities in terms of legible design, maximal control, and efficient allocation, they would also become spaces of endless eavesdropping.

The politics of public listening have always been complex, and the architects of smart cities offer little to intensify their disparities. As Cranshaw argues, "If an individual does not want to be sensed by the distributed sensors throughout the city, does she have that choice? If she does not control any of the devices sensing her, opting out seems hopeless."[22] Unlike traditional experimental ethics, where informed consent, opt-outs, and other accountability mechanisms are involved, the model of the urban laboratory appears to have very few rules when it comes to how citizens will be included. Sally Applin has described how systems of modern life are increasingly designed to require us to use data systems, a process she calls "forced compliance," particularly in a labor context.[23] But now there is a significant slippage between the data systems that we elect to use, be it a smartphone or Facebook, and the systems that physically surround us. In a big data planned space, you are being tracked and observed. When this watching is combined with predictive modeling techniques, you exist as data regardless of whether or not you carry a cell phone or use the Internet at all.

Thus, in many of these big data experiments, simulations, and "pilot tests," citizens are in a state of *forced participation*—existing as single data points within a regime of endless measurement, driven by the desire to unearth all hidden places and thoughts, and to illuminate all darkened corners.

This problem is further compounded by the economic rhetoric of mass data collection: that it will lower the costs of insurance, transportation, and policing while simultaneously increasing safety and efficiency. As Bohn et al. observe, "Given the immediate economic returns of consumer loyalty programs or low insurance rates, the rather vague threats of future privacy violations are easily ignored."[24]

Michel de Certeau once described his image of a horrific city as an "urban text without obscurities, which is created by a technocratic power everywhere and which puts the city-dweller under control (under the control of what? No one knows)." He describes spaces "brutally lit by an alien reason," immobilizing society into a transparent text.[25] In short, he saw a society "characterized by a cancerous growth of vision, measuring everything by its ability to show or be shown . . . a sort of epic of the eye."[26] While this is a vivid account of the totalizing force of an instrumented city, it would seem that beyond Certeau's metaphors of vision, big data planning can be understood as a practice of algorithmic listening—listening without limits and without consent. Where Certeau suggested the twentieth-century city could be understood as an epic of the eye, the metaphor for ongoing ever-present data collection in our century has become an all-receiving listening ear.

The idea of "listening in" has been central to the last fifty years of legal thinking about privacy. In 1967, the canonical example of a privacy breach was the state listening to a man in a telephone booth. In *Katz v. United States,* a man talking on a phone was recorded by FBI agents who had attached an electronic listening device to the outside of the public telephone booth. The U.S. Supreme Court ruled that even though the government argued that the telephone booth was glass, and therefore not a private space, what the man "sought to exclude when he entered the booth was not the intruding eye—it was the uninvited ear."[27] It seems somewhat extraordinary now to reflect on how bugging a telephone booth was so recently deemed serious overreach and became the foundation of a "reasonable expectation" of privacy. Since 2013, with the release of intelligence agency documents by Edward Snowden, we know that we live in an era where the NSA can collect all forms of network data. Yet the NSA continues to make the paradoxical claims that they are "gathering data but not collecting it"— because in the agency parlance, collection only occurs when they begin to actively process data. In this model, listening is constant, but it only triggers certain legal obligations and safeguards once they begin tuning in to a specific channel, be it in relation to an individual or a group.

Several years ago, I wrote about the concept of listening as a metaphor for the forms of participation online that are commonly overlooked or entirely discounted.[28] Listening defines the more receptive modes of online engagement which occupy the vast majority of our time: reading the work of others and favoriting and liking—but not necessarily commenting or posting. And while "lurking" tends to be used pejoratively, it is a necessary part of how online communities form: People come back to listen. The tendency of Internet studies has been to celebrate "voice" as the sine qua non of participation, and to place great value on "speaking up" in online spaces of debate such as blogs, social media, and news sites. But the hierarchical understanding of participation that values speaking over listening ignores the agentic power of listening and the way it contributes to social intimacy and connectedness.[29] Listening is an important type of participation—and my earlier work was interested in how this mode has its own values, norms, and disciplines.

But there is now a significant slippage between the digital and the physical: Where people are aware that there are many agents listening in environments like social media platforms and blogs, there is less awareness of the ways in which routine activities like walking down the street have themselves become legitimate data sets to be captured and analyzed. This is matched with a profound imbalance in the ability to listen in— states and corporations (both separately and together) are listening to citizens as *adversaries*. The result is a radical disparity where those with the greatest economic and technical power are turning their strengths against those with the least. This is a unidirectional "listening in" rather than any kind of reciprocal, intersubjective experience of shared listening. The listening that is now occurring—whether it is done by the city of Boston, the NSA, or Facebook—is so pervasive that its true power is only visible when seen in the aggregate.

From the Statistical City to the Simulated City

If the model of the purpose-built smart city is yet to be realized, where else can we look to understand the current model of data-driven simulations in the built environment, where experimental "lab" rhetoric is being applied to urban space?

Gravesend is a London suburb that contains within it a shadow, simulated suburb: a village custom-built in 2003 by the Metropolitan Police, working alongside two private firms called Equion Facilities Management and Advanced Interactive Systems. Together they have created a training ground that initially looks like a suburb, but the streets are empty and the buildings are merely facades. It is an immersive set designed to simulate the real lived environment: Here police stage all manner of events, from riots and hostage scenarios to bank robberies and mass protests. Using Google Earth, you can see the scale of the simulation, which is otherwise closed to the public. Chris Clarke, who maintains a photo series of Gravesend, describes it as an urban facsimile: "This surreal installation serves as a chilling account of the death of community in 21st century Britain."[30] Architecture writer Geoff Manaugh calls it "the civic minimum."[31]

Gravesend, of course, is just one of many private military training arenas around the world. We could just as easily consider the Urban Warfare Training Center in Israel's Tze'elim military base in the Negev Desert—a massive "mock city" that is heavily equipped with cameras, sensors, and an audio system that simulates helicopters, mortar rounds, and prayer calls. These zones of exception were first designed as a way to test out scenarios, train personnel, and understand a terrain—be it urban or otherwise.[32] Understandably, this had to be done away from the normal built environment due to the risk to citizens, and it was understood not to be the stuff of everyday life. The normal civic rules are suspended in these simulations: They do not have to apply for permits, or seek approval from local councils, or follow local laws. Like a permanent space of martial law, these are simulations of cities where military authority prevails over any normal administration of law and democratic functioning.

But now there are forms of simulation that do not require the creation of a separate city, with all the work and expense of infrastructural engineering. Big urban data programs allow for forms of information gathering and analysis and also offer fertile grounding for urban experimentation, enabling the development of models and the running of simulations on the data gathered from a city's inhabitants, generally without their consent or even an awareness that they are participants. By using real urban environments, rather than mock cities or software simulations, the claim can be made that the resulting analysis is both "real" (using actual cities, with their own peculiarities and characteristics) and "scientific" (experiments where city agencies or commercial ventures can modify conditions and see how communities respond). But although the data collected in cities may be "real," the models deployed, and the predictions drawn from them, are virtual; they are abstractions, but they may have real effects. In this sense, they are simulations, imitations of real-world systems that seek to produce knowledge about the cities they seek to model.

This explains the common use of the laboratory metaphor by big urban data programs such as the Center for Urban Science and Progress (CUSP), which describes itself as "a public-private research center that uses New York as its laboratory."[33] In CUSP's promotional literature on its website, we can see the rationale laid out in terms of data as providing a form of scientific ground truth, while also offering affordable simulation possibilities. They argue that their focus is on data rather than "more physical forms of engineering" because "it is impossible to judge the efficacy of steps to improve a city without well-defined metrics and baselines" and that "data-only research is far less capital-intensive than physical activities, which require expensive specialized laboratories."[34] Their program includes pointing a high-resolution camera at apartments in New York to track when people turn off their lights—more as a kind of experiment or pilot rather than addressing a particular need or problem. "It's like when Galileo first turned the telescope on the heavens," the director of CUSP, Steve Koonin, told the *Wall Street Journal*. "It's just a whole new way of looking at society."[35]

Here we see the confluence of two logics: the "statistical city" in Jacobs's terms, and the "simulated city." This combination is particularly appealing to those who would seek both a God's-eye view of an urban space, and those who would like to stage particular kinds of tests (scientific or otherwise) and trial potential interventions. According to CUSP, the concept of the urban laboratory holds the strongest appeal for four groups: government, the private sector, security organizations, and social scientists. And while there is a long history of the first three groups seeking to extract maximal data, legibility, and value from populations, the inclusion of social scientists remains controversial.

Should social scientists be extracting data and staging experiments on communities that are unaware, and possibly unconsenting? This cuts directly to questions of research ethics, as well as to the traditional understanding of the scientific method. Leah Miesterlin argues that both are at risk—in her essay "The City Is Not a Lab" she critiques the notion of cities as sites for social experimentation. Cities, she notes, are dynamic spaces that fail to produce the conditions of controlled inquiry. The new techniques of urban data gathering are premised on the belief that more data will "flesh out our models such that they come to represent (rather than abstract) the city itself"—a case of confusing the map for the territory. Instead, the turn to big data is producing analytical shortcomings that are leading to false knowledge claims and unforeseen human effects when that research is applied to urban contexts.

> As our ability to model complex systems grows, augmented by the ubiquity of data collection and the increase in urban empiricism, so too grows the data-driven and computational functioning of the city itself. This parallel development has led to a confusion of methods and research framing, particularly surrounding urban informatics and the use of data visualization in architectural inquiry.[36]

To put it another way, gathering data has become both method and justification, before a research question has even been determined. It has resulted in claims that urban experiments are produced, despite the basic requirements of scientific method not being met. Tests are not replicable, as architects and planners "can never fully 'reset the experiment,' whether to undo or to verify results."[37] Urban tests change the system as well as the conditions being tested, losing the possibility of replication. Moreover, this approach amounts to "experimenting on human subjects without their consent . . . and without appropriate measures of accountability for the impacts levied upon those populations."[38] Simply put, it should not pass an ethics review until we have a far more developed sense of what kinds of ethical parameters should be placed around this kind of big urban data work.

What we are seeing is the encroachment of the politics of simulation into lived urban space. Gravesend and other simulated spaces like it are contained nonplaces where rules can be suspended and controllable experiments can be staged, but now big data techniques and tracking infrastructures are bringing those logics back home. They offer the ability to overlay simulations across an entire city, or parts of it, at a time, to seek new forms of information and to test new interventions. But as Miesterlin observes, this is not reproducible science. Nor is it a recognizable form of democratic participation in lived environments—in what Jacobs calls the organized complexity of cities. Instead, it is much closer to the civic minimum.

Conclusion

Data is now seen as enormously valuable in the understanding and development of cities, and thus there are strong justifications for extracting it wherever possible. Further, in the context of ongoing crises—be they natural disasters or terrorism—the constant monitoring and analysis of urban populations is commonly framed as essential in meeting these challenges. Thus tracking systems that extract sensitive data from citizens are being widely trialed and tested without the citizens' knowledge or consent, harvesting location data and information that points to markers of race, gender, and class. From the perspective of the statistical city, these data will always embody useful knowledge—even though the systems that gather the data cannot promise its representativeness. Given that the history of massive data sets and statistical urbanism is so recent, we are just beginning to see the full impact of how these approaches will change the ways cities function. But the last hundred years should suggest caution, as Jacobs argued over fifty years ago.

The current historical moment of big urban data emphasizes remote experimentation and autonomous simulation—yet there is little evidence of systemic improvement resulting from this nascent set of practices. Certainly the last decade of urban pilot tests and "city as laboratory" programs has brought serious questions to the forefront: What roles do citizens have in shaping the future of cities? Who gets the God View? What does participation mean in a smart city? If the power represented in the systems that listen to the city is concentrated in city offices and commercial data firms, we have reason to question whether these are systems that support the democratic functioning of civic space. As Joerges argues, "Built spaces always represent control rights. They belong to someone and not to others, they can legitimately be used by some and not by others."[39] In order to judge whether the cities of the future are participatory for anyone other than the most powerful, we have to ask who is listening, how they listen, and whether those social categories can be contested.

Notes

1. Details come from an investigation conducted by the local Boston publication *Dig Boston*. Chris Faraone, Kenneth Lipp, and Jonathan Riley, "Boston Trolling (Part I): You Partied Hard at Boston Calling and There's Facial Recognition Data to Prove It," *DigBoston*, August 7, 2014, http://digboston.com/boston-news-opinions/2014/08/boston-trolling-part-i-you-partied-hard-at-boston-calling-and-theres-facial-recognition-data-to-prove-it/.

2. Ibid.

3. Ibid.

4. Don DeLillo, *White Noise* (London: Picador, 1985), 139.

5. See Bloomberg's statement at the Mayor's Challenge, March 13, 2013, http://mayorschallenge.bloomberg.org/index.cfm?objectid=D4C15F70–8BCF-11E2-AF4B000C29C7CA2F.

6. Orit Halpern, Jesse LeCavalier, Nerea Calvillo, and Wolfgang Pietsch, "Test-Bed Urbanism," *Public Culture* 25 (2013): 279–80, doi: 10.1215/08992363–2020602.

7. See Andrés Monroy-Hernández, Shelly Farnham, Emre Kiciman, Scott Counts, and Munmun De Choudhury, "Smart Societies: From Citizens as Sensors to Collective Action," *Interactions* 20 (2013), doi: 10.1145/2486227.2486249; Sallie Ann Keller, Steven E. Koonin, and Stephanie Shipp, "Big Data and City Living," *Significance* 9, no. 4 (2012): 4–7; Gary King, "Ensuring the Data-Rich Future of the Social Sciences," *Science* 331 (2011), http://gking.harvard.edu/files/datarich.pdf/.

8. Adam Greenfield, *Against the Smart City* (New York: Do Projects, 2013), Kindle edition, chapter 4.

9. Ibid., chapter 12.

10. Tiffany Winman, "IMB Goes to Washington to Fight Terrorism," *IBM Blog*, June 16, 2008, https://www-304.ibm.com/connections/blogs/tiffany/entry/ibm_goes_to_washington_to_fight_terrorism?lang=en_us.

11. Ibid., 63.

12. Robert Caro, *The Power Broker: Robert Moses and the Fall of New York* (New York: Knopf, 1974).

13. See Langdon Winner, "Do Artifacts Have Politics?" *Daedalus* 109 (1980), and Bernward Joerges, "Do Politics Have Artefacts," *Social Studies of Science* 29 (1999), doi: 10.1177/030631299029003004.

14. Jane Jacobs, *The Death and Life of Great American Cities* (New York: Random House, 1961), 435.

15. Ibid., 436.

16. Ibid., 436–37.

17. Ibid., 437.

18. Ibid., 438.

19. Ibid., 436.

20. Kashmir Hill, "'God View': Uber Allegedly Stalked Users for Party-Goers' Viewing Pleasure," *Forbes*, October 3, 2014, http://www.forbes.com/sites/kashmirhill/2014/10/03/god-view-uber-allegedly-stalked-users-for-party-goers-viewing-pleasure/.

21. Winman, "IBM Goes to Washington."

22. Justin Cranshaw, "Whose 'City of Tomorrow' Is It? On Urban Computing, Utopianism, and Ethics," paper presented at the 3rd International Workshop on Urban Computing, Chicago, Ill., August 11–14, 2013, 4.

23. Sally Applin and Michael Fischer, "Forced Compliance: How the City Shapes the Network That Shapes the City," paper presented at the joint conference of the Japanese Society of Cultural Anthropology and the International Union of Anthropological and Ethnological Sciences, Chiba City, Japan, May 15–18, 2014.

24. Jürgen Bohn, Vlad Coroamă, Marc Langheinrich, Friedemann Mattern, and Michael Rohs, "Living in a World of Smart Everyday Objects — Social, Economic, and Ethical Implications," *Human and Ecological Risk Assessment* 10 (2004): 771, doi: 10.1080/10807030490513793.

25. Michel de Certeau, *The Practice of Everyday Life,* trans. Steven Randall (Oakland: University of California Press, 1984), 94.

26. Ibid., xxi.

27. *Katz v. United States,* 389 U.S. 347 (1967).

28. Kate Crawford, "Following You: Disciplines of Listening in Social Media," *Continuum* 23 (2009), doi: 10.1080/10304310903003270.

29. Kate Crawford, "Listening, Not Lurking: The Neglected Form of Participation," in *Cultures of Participation: Media Practices, Politics and Literacy,* ed. Hajo Greif et al. (Berlin: Peter Lang, 2011), 66.

30. Chris Clarke, "Gravesend — The Death of Community," https://www.flickr.com/photos/chris-clarke/sets/72157627388065874.

31. Geoff Manaugh, "The Civic Minimum," *BldgBlog,* October 18, 2014, http:// bldgblog.blogspot.com/2014/10/the-civic-minimum.html.

32. James Der Derian, *Virtuous War: Mapping the Military-Industrial-Media-Entertainment Network* (Boulder, Colo.: Westview Press, 2001).

33. Center for Urban Science + Progress, http://cusp.nyu.edu/.

34. Steven Koonan, "The Promise of Urban Informatics," Center for Urban Science + Progress, last modified May 30, 2013, http://cusp.nyu.edu/wp-content /uploads/2013/07/CUSP-overview-May-30–2013.pdf.

35. Elizabeth Dwoskin, "They're Tracking When You Turn Off the Lights," *The Wall Street Journal,* October 20, 2014, http://online.wsj.com/articles/theyre -tracking-when-you-turn-off-the-lights-1413854422.

36. Leah Meisterlin, "The City Is Not a Lab," *Applied Research Practices in Architecture Journal* 1 (2014), http://arpajournal.gsapp.org/the-city-is-not-a-lab/.

37. Ibid.

38. Ibid.

39. Joerges, "Do Politics Have Artefacts?"

CHAPTER 10

The Pacification of Interactivity

Mark Andrejevic

O NCE UPON A TIME, not long ago, music served as the paradigmatic example of the disruptive potential of digital media technologies. At the turn of the new millennium, file sharing came to stand as the paradigmatic example of emerging forms of bottom-up participation that subverted the top-down, market-driven tactics of marketers, producers, and the culture industry more broadly. Product managers at Microsoft, working away on software to play MP3 files, proclaimed, for example, that "We really are in the middle of a digital media revolution. . . . We have learned that people are not just using their computers for productivity."[1] Participating meant more than uploading comments or putting up a personal Web page—it meant taking part in the circulation of music and in so doing literally taking partial control of content that had previously been more tightly controlled by commercial entities, and, in so doing, taking apart an existing business model. This type of participation was disruptive, perhaps subversive, sometimes illegal, but above all active—that is, it required some degree of time, effort, and intentionality on the part of the participant. However, alongside the new and retooled forms of activity facilitated by digital media technologies were new forms of data capture enabled by the emerging sharing networks and platforms. Thus, for example, the company BigChampagne capitalized on the information generated by file-sharing networks to monitor music popularity. Even as the recording industry mounted legal challenges to file-sharing networks, it exploited them in the interest of creating the "the world's biggest focus group."[2]

The lesson of this conjunction of surveillance and subversion is that, in the digital era, active participation generates data about itself in addition to the intentional and deliberate forms of action or feedback with which it is associated. This is a crucial—and increasingly important—aspect of the digitally informed society: Digital participation is reflexive in the sense that it generates information about itself, and this information may be more detailed and comprehensive than the information generated by deliberate and active forms of participation. If, then, we are to describe participation as a form of contribution to a range of ongoing processes, it is crucial to note that *active* participation can be redoubled in its passive form. The essay explores both the trajectory of the passive-ication (and pacification) of participation via interactivity, and some of its implications for considering the forms of management and influence envisioned by those who control and mine the booming digital databases. Henry Jenkins suggests a distinction between participation and interactivity that assigns the former to the realm of social practice and the latter to that of technological affordance. Thus: "Interactivity refers to the ways that new technologies have been designed to be more responsive to consumer [or user] feedback. . . . Participation, on the other hand, is shaped by the cultural and social protocols. . . . Participation is more open-ended, less under the control of media producers and more under the control of media consumers" (133). For Jenkins the distinction is between technological and social protocols—technical versus social code—with the understanding that the latter is, in many contexts, more open to transformation by a user or groups of users. Perhaps understandably, even this distinction blurs in certain contexts: Technologies that allow great opportunities for feedback (a blog versus a traditional newspaper, for example) simultaneously facilitate new forms of participation—and thus impact social protocols. This chapter argues for pushing the distinction further: Participation carries connotations of collaborative construction of the protocols themselves. Technically, participation simply means "to take part" in: I might participate in a survey, for example. However, there is an underlying promise connoted in the offer to participate. Framed in these terms, the invitation to participate carries with it the implication that this activity will be meaningful for the user, a form of self-expression and, simultaneously, a participation in the greater good. To take an extreme example, we might describe a guided missile as having an interactive relationship with its target (insofar as the target provides feedback to the missile) but not a participatory one (unless the target is suicidal): To say that the target *interacts* does not mean that it participates in its own destruction. Interactivity takes place according to coded protocols; participation carries the implied promise of intervening in the code. This chapter, then, is concerned with the ways in which participation is redoubled in the form of interactivity when it generates information about itself that can be construed as feedback but not collaboration. The recurring ideological move of the interactive economy is to bulldoze this distinction, equating the provision of feedback with participation—not least because of the alleged convenience of the automation of participation. Interactivity offers a possible solution to the pressures created by the injunction to participate: Now! More! Faster! (Tweet! Post! Update! Blog!).

The vicissitudes of the music industry provide one possible locus for what this paper describes as the passive-ication of interactivity. Consider, for example, Facebook's development of an app to capture information about users' listening habits when they post status updates from their mobile devices. The utility will automatically activate microphones on users' devices to catch a short burst of audio that will then be matched to a database of music and television content to determine what the user is watching or listening to so that this information can be automatically shared with friends (and, presumably, interested marketers).[3] Facebook claims the audio will be deleted once the match is made and that it is not storing actual audio captured from user devices (which would presumably include all sounds the microphone might pick up)—but of course, it is still capturing the details of users' listening and viewing patterns throughout the course of the day. In this regard, the application might be considered a distant descendant of the early telephone "coincidentals" that transformed the radio ratings industry in the 1930s. Instead of asking people (who had agreed to *participate*) what music they listened to the day before, ratings industry innovator C. E. Hooper pioneered the method of calling up during various times of the day and asking people what they were listening to at the time (if anything). The virtue of such an approach was not simply that it didn't rely on recall (or on some perceived sense of what people thought they *should* have been listening to), but also that it captured the rhythms and patterns of music consumption through the course of the day. In the case of the Facebook app, even the intrusion (and cost) of the telephone call is eliminated: The app automatically captures and analyzes the data whenever a user posts a status update. The more one posts, the more comprehensive a profile of one's viewing habits. For the user, the apparent benefit is the ease of sharing information about viewing and listening practices with "friends." For Facebook, however, there are a range of benefits: not simply further data about user preferences and practices, but also indications of the popularity of music and TV content with particular users, as well as an increasingly comprehensive portrait of the flow of viewing and listening habits—not to mention the stimulation of users' engagement with the site: The more one shares, the more likely one is to elicit responses from friends online.

As is often the case in the convergent realm of data analysis, capturing information about listening behavior is about much more than music sales. Thanks to the fact that more people are carrying microphones with them wherever they go (embedded in their phones), the ability to capture information about listening habits has expanded to cover a greater range of populations and practices than ever before. Unsurprisingly, then, companies like Shazam and Facebook are staking out a rapidly expanding frontier for digital-era commodification and brand capture. The giant media-services agency Mindshare is interested in using audio data the way other companies use Web and image data: "We spend a lot of time with Google optimizing keywords for search, and a lot of time with Instagram and Pinterest talking about what brands should be doing with imagery, but not a lot of time on sound. . . . It's the sonic territory that the brand would like to own."[4] Ownership, in this regard, means finding ways to put the data to use to track and profile consumers, to monitor the circulation of viewing and listening practices, and also to use the resulting data for as wide a range of useful and productive inferences as possible, ranging across the realms of employment, politics, security and policing, and perhaps even health care and education. As the founder of the music data company The Echo Nest (which develops systems for categorizing and grouping different types of music) put it, "We've been collecting this data for a while now and started looking into what correlations exist between music, psychographics, demographics and other media preferences."[5] The music streaming company Pandora has already indicated that it mines listener data to infer the political preferences of listeners—with an eye toward using this information for political marketing.[6]

The broader point to be made is that the amount and type of data that goes into this kind of analysis and sorting far surpasses the information consciously and intentionally provided by users. Relatively early on in the interactive era of digital communication, users were asked to do the work of describing themselves to one another—thereby creating a pool of useful information about themselves for one another and for the platforms they were using. Thus, for example, Facebook asked users to provide descriptions of their tastes in music and Facebook Groups further formed the basis for self-categorization and self-disclosure. Sites like MySpace and Friendster similarly exploited the intersection of self-expression, self-disclosure, self-display, and self-categorization. However, as the data-driven economy ramped up, it became clear that for the purposes of sophisticated forms of profiling and customization, more data could provide higher degrees of specification and accuracy. Data scientist Alex Pentland has discovered, for example, that mobile phone usage information can indicate when people are falling ill before they realize it themselves.[7] Researchers working for Microsoft have developed mood recognition software that they claim can accurately assess a user's mood 93 percent of the time based on patterns of mobile phone use.[8]

The list goes on: The productivity of inferential monitoring is directly related to the exploratory character of data mining, which promises to unearth unanticipatable and unintuitable patterns based on the fact that variables one might not expect or imagine to be related can generate useful insights. For example, the Internet browser one uses might turn out to be a good indication of future job performance, where one went to school might correlate with what brand of razor one prefers, and so on.[9]

Inferential monitoring is both speculative and indefinitely productive: There are no a priori limits on which variables might generate useful correlations. Unearthing these correlations, however, requires increasingly comprehensive and rich data profiles. Intentionally provided and deliberately shared data falls far short of meeting the demands of this kind of data mining, in part because it is time consuming to consciously and intentionally provide data (to fill out one's preferences; to document one's activity; to characterize one's own tastes, behavior, activities; and so on), and in part because self-profiling introduces the element of self-consciousness and thus the potential for self-censorship or artifice. Alex Pentland's admonition to data scientists to track what people *do* rather than what they say conserves the familiar (though not uncontested) social scientific notion that *what is done* (as opposed to what is said) is likely to offer—in some sense—more authentic, more factual, more valuable data because it is less subject to the vagaries of self-conscious manipulation and less susceptible to falsehood.[10]

The virtue, then, of passive data collection is that, like hidden cameras, it supposedly eliminates the element of reflexivity—that is, the ways in which the monitoring process, once detected, might transform what is being monitored. The pacification of "interactivity," then, has the following advantages over self-report (*active* information generation) from the perspective of the data collectors: (1) It overcomes the limits on quantity of data collected; (2) it skirts the hazards of observer effects by rendering data collection less obtrusive; and (3) it opens up space for the indefinite propagation of new dimensions of monitoring. In purely practical terms, passive data collection does not rely on practices that would alert individuals to the data collection process. Thus, for example, people who are listening to Pandora have no way of knowing that their political preferences are being associated with their listening habits; people sharing files online are not necessarily aware that each time they download a song, they are contributing to a geolocational ratings system; people driving down the street aren't aware that their mobility patterns are recorded by their mobile phones (or their automatic toll passes, or an array of automated license plate readers).

Agency Migration

The obverse of the passive-ication of interactivity from the point of view of the monitored subject is the hyperactivity of the monitoring apparatus. As the need for individual participation (via the active, self-conscious process of filling out forms, posting preferences and likes, cataloguing tastes, and so on) is curtailed, the "active" capacity of the monitoring apparatus is enhanced. The Facebook user no longer has to type in information about what he or she is watching or listening to because the device has developed new data-capture capabilities. The user still has to take action to trigger the music-sharing app—but the underlying tendency is toward complete automation of interaction.[11] Thus, for example, users of SoundCloud can use a link to Facebook to automatically share their listening habits: "If you enabled sharing sounds you listened to on SoundCloud to Facebook, all sounds you listen to will be posted to the Facebook Ticker, no matter where you're listening on the web. So that means if you listen on the iPhone, iPad, Android app, mobile site, and/or via an embedded player while logged into SoundCloud, these sounds will be shared to Facebook."[12] Perhaps this shift toward the automation of sharing is best summed up by the tag line for a dating app (designed for Google Glass) that uses facial recognition technology to identify potential matches (and to screen out undesirables): "With NameTag, Your Photo Shares You."[13] As the app's promotional literature puts it, "NameTag can spot a face using Google Glass' camera, send it wirelessly to a server, compare it to millions of records and in seconds return a match complete with a name, additional photos and social media profiles."[14] The app developer, FacialNetwork.com, is also developing technology to match photos with criminal databases, including the U.S. National Sex Offender Registry.

The logic of the app's tagline can be adapted to describe other passive monitoring systems: "With Pandora, your music shares your political preferences"; "With MoodScope [an app developed by Microsoft researchers], your phone shares your mood"; "With GPS, your car shares your location"; and so on. What emerges is a kind of phantasmagoria of objects in constant communication with one another—tracking, processing, analyzing, and responding in highly complex networks of interaction. From a familiar critical perspective, such a phantasmagoria has overtones of fetishization, the displacement of human relations onto the object world—a kind of digital upgrading of the commodity spectacles of the nineteenth century, in which "the living, human capacity for change and infinite variation becomes alienated and is affirmed only as a quality of the inorganic object."[15] The upgrade is necessary to address the sheer demand for data: The amount of data collected on a regular basis surpasses both the ability of human hands to supply and collect it, and of human minds to comprehend it.

The apparently urgent desire for hyperactive objects, embedded in discourses about the "Internet of things," and the proliferation of "smart" devices and places (cities, homes, schools, and so on), highlights a significant shift in the valence of the process formerly known as fetishization. In its critical, Marxist conceptualization, the notion of fetishization refers to a confounding conceptual error: the misrecognition of social relations between people once these relations become absorbed into the commodity form. In Volume 1 of *Capital*, Marx explains his conception of the relationship between the commodity form and religion: "In that world the productions of the human brain appear as independent beings endowed with life, and entering into relation both with one another and the human race. So it is in the world of commodities with the products of men's hands. This I call the Fetishism which attaches itself to the products of labour, so soon as they are produced as commodities."[16]

As Slavoj Žižek emphasizes, the goal of critique associated with this conception of fetishism is not simply demystification—that is, the direct reversion to the underlying source of the fetish—but rather the attempt to "unearth the 'metaphysical subtleties and theological niceties' in what appears at first sight just an ordinary object. Commodity fetishism (our belief that commodities are magic objects, endowed with an inherent metaphysical power) is not located in our mind, in the way we (mis)perceive reality, but in our social reality itself."[17] Any straightforward attempt at demystification bypasses the important engagement with the social processes whereby, "In your social reality, by means of your participation in social exchange, you bear witness to the uncanny fact that a commodity really appears to you as a magical object endowed with special powers."[18] In a reflexively savvy era, a familiar contemporary response to capitalism is to "see through" the mythology of the "free" market (and the attendant ideologies of "fair" exchange, accurate allocation of value, and so on), and yet this persists despite its savvy debunking, because it depends not so much on abstract beliefs as on the embodied social practices they once supported. If the practices can continue unabated in the face of the dissolution or debunking of their ideological alibi, they take on a certain uncanny force.

Perhaps unsurprisingly, then, forms of postideological critique look beyond discourse to the realm of rehabilitated materiality.[19] Intriguingly, the reconfiguration of materiality in some versions of "new materialism" repositions the process previously known as "fetishization." The rehabilitation of fetishization can exhibit an inadvertent affinity to the enthusiasm for the phantasmagoria envisioned by emerging forms of passive interactivity. This repositioning of the figure formerly known as the fetish is suggestive when placed against the background of proliferating "smart" objects and increasingly autonomous interactive devices. Consider, for example, the "vibrant" materialism of Jane Bennett, which seeks to restore to the object realm the vitality suppressed by an overweening anthropocentrism.[20] From such a recuperative perspective, the attempt to reduce the animation of matter to the simple misrecognition of a displaced human agency carries with it overtones of a reductive narcissism. In her earlier work on enchantment, Bennett critiques Marx's notion of commodity fetishism for its tendency "to picture a matter whose 'natural character' is dead or inert: for him nature is dis-enchanted. That is to say, it is drawn against the background of a superseded pagan world wherein all things were enlivened with divine spirit. In doing so, it supports an onto-story in which agency is concentrated in humans."[21]

By way of contrast, Bennett uses the example of a Gap television ad for khaki pants, in which "the khakis dance along with the human bodies in them," to capture the vitality of a reanimated object world. As she puts it, "I see the GAP ad as expressive of a different ontology: in it, the liveliness of matter itself is once again apparent, this time by the grace of cinematic technology rather than God or the Spirits."[22] We might add to this formulation another two graces: those of digital animation and interactive technology. In the era of smart clothes and consumer robotics, the pants may not even need the Hollywood touch to come alive. In fairness, Bennett's political commitments are to ecology and social justice, but where Marx discerned only alienation and misrecognition in the animation of the commodity world associated with capitalism, she discovers a nascent political potential, both in the apparent reenchantment of the object world and in the tempering of a destructive anthropocentrism. As Bennett puts it, "Even today, the electricity generated by running on the treadmill of consumption is not a closed circuit, and those charges are also powering other, non-commercial phenomena."[23]

Indeed, it is not hard to discern in Bennett's account a rehabilitation of this mystical animation—and the attempt to turn it to purposes other than consumption for its own sake: "You can call those pants 'commodified' and you can call fascination with the advertisement a 'fetish,' but the swinging khakis also emerge from an underground cultural sense of nature as alive, as never having been disenchanted. Out of the commercialized dance erupts a kind of neo-pagan or Epicurean materialism. An enchanted materialism."[24] The dancing khakis, she asserts, display "a playful and surprising will."[25] The importance of this "will," for Bennett, is the space of possibilities opened up by the inability to reduce it entirely to the dictates of capitalism, commercialism, and commodification: "I aim instead to deny capitalism quite the degree of efficacy and totalizing power that its critics (and defenders) sometimes attribute to it, and to exploit the positive ethical potential secreted within some of its elements."[26]

The risk of such an approach—which Žižek has described as "weak pan-psychism or terrestrial animism"—lies in its inadvertent homology with the re-animation of the object world and the dispersion of agency envisioned by the passive-ication and automation of interactivity."[27] It is also worth asking what might get lost in the process of reenchantment, insofar as it circumvents the critical interpretive move associated with the critique of fetishization: a consideration of the social relations that make possible a particular way of thinking and knowing. The loss is a significant one, because it tends to reinforce a lack of reflexivity regarding the ways in which a mode of thinking is embedded within its own historical context. More pointedly, it forestalls the attempt to address the question, raised by Alex Galloway, about the complicity of recent theoretical approaches with economic processes: "Why, within the current renaissance of research in continental philosophy, is there a coincidence between the structure of ontological systems and the structure of the most highly evolved technologies of post-Fordist capitalism?"[28] Galloway is referring here to the alignment between object-oriented programming and so-called object-oriented ontology, but notes that this connection has wider resonances. We might point, for example, to the way in which the dispersion of agency associated with variants of new materialism mimics and embraces the indefinitely expanding logic of the database, in which a growing range of variables are brought to bear in predicting particular outcomes. The database treats all of its inputs—human and otherwise—as equal participants in pattern generation. Thus, for example, crime-detection algorithms will compile past crime data alongside meteorological data and information about building code violations. In the great "democracy of things" represented by the database, data about air pressure, broken windows, and the number of daylight hours commingle with the more distinctly human statistics about poverty and abuse. The significance of the inputs can only be assessed, as it were, after the fact. However, all are endowed with a role in the construction of potentially useful patterns from the data and thus are treated as contributors to particular outcomes, whether these be crime rates, purchase volumes, catastrophic events, the circulation of diseases, or otherwise.

The result of this indefinite dispersion of agency is that the "agent" becomes the *outcome* and not the cause of a particular "assemblage" (*agencement* in the original French). From this perspective, narratives of causality that rely on a necessarily delimited range of agents end up being too simplistic—there are many more variables in play than such accounts allow. Multiple variables interact in a configuration that is too complex to be translated into narratives of intent and causality. Rather, a particular outcome must be understood as the result of an endlessly unfolding series of interactions that would, in the end, be futile to enumerate. The corollary is that subjective notions of intentionality are subsumed by the assemblage: "an intention is like a pebble thrown into a pond or an electrical current sent through a wire or network: it vibrates and merges with other currents, to affect and be affected."[29] Such a formulation does not simply disperse or distribute agency, it also reassembles it in the form of the "agentic capacity" of the assemblage—what Bennett describes as "the dynamic force emanating from a spatio-temporal configuration rather than from any particular element in it."[30] The agent is the outcome; or, transposed into the realm of communication, the message is the effect, as Jeremy Packer has argued in his discussion of data mining. He draws on the example of Google, whose "computations are not content-oriented in the manner that advertising agencies or critical scholars are. Rather, the effect is the content. The only thing that matters are effects—did someone initiate financial data flows, spend time, consume, click, or conform? Further, the only measurable quantity is digital data. Google doesn't and couldn't measure ideology."[31]

The demotion of subjectivity and intentionality, in their distinctively human guise, characterizes developments in both theory and the emerging practices associated with "big data" mining, highlighting what Galloway describes as "the coincidence between today's ontologies and the software of big business."[32] The two converge, in a sense, on a reconfiguration of those forms of activity and participation associated with subjectivity and intentionality. In both cases there is a focus on the logic of emergence. Big data mining is portrayed as an emergent process in the sense that the most useful correlations are unanticipatable: The only way to generate them is to run the process. Similarly, material interactions are emergent in the sense of being not fully predictable or determinist, thanks to the asubjective "agency" of matter, which refers generally to "an efficacy that defies human will."[33]

The reenchantment of the recalcitrance of the object world fills in the space for critique opened up by the concept of fetishization. The symptoms of such a postcritical stance abound in variants of new materialism, object-oriented ontology, and even quite different approaches, such as that associated with German media theory—what Packer describes as "the new Germans."[34] In opening up the field of explanation to a potentially infinite chain of agentic factors and forces, Bennett's work on vibrant matter shares a certain postnarrative tendency with these quite disparate approaches (once narratives multiply indefinitely, they neutralize themselves)—and it is this shared logic that is the focus of the current analysis.

More generally, one of the effects of the ostensibly democratic form of deanthropocentric leveling and the generalization of agency is a dedifferentiation of actors, priorities, and relationships that poses challenges for politically motivated forms of critique. Jorgensen's (2012) sympathetic but critical analysis of Karen Barad—whose work has been influential in the development of new materialist approaches—lays out some of the key challenges raised by theoretical approaches that resist privileging any particular set of actors so as to reconfigure notions of agency and intention. Such an approach, she argues, "is wanting in terms of political potential—almost to the extent that political action becomes unfeasible . . . capacity for action cannot be localised, and it thus becomes unclear where and how political action can be brought about."[35]

The agency of the "assemblage" threatens to gloss over the internal conflicts and contradictions that are subsumed in the eventual outcome—the moment at which agency emerges, as it were, after the fact. It is perhaps not surprising, then, that critique may not have an important role to play in new materialist accounts. As Barad puts it, "I am not interested in critique. . . . Critique is all too often not a deconstructive practice, that is, a practice of reading for the constitutive exclusions of those ideas we can not do without, but a destructive practice meant to dismiss, to turn aside, to put someone or something down—another scholar, another feminist, a discipline, an approach, et cetera."[36] Absent from this formulation is the version of critique associated with critical theory—neither deconstruction nor one-upmanship, but an exploration of the social conditions that foster immiseration in an effort to transform them. Critique, in this sense, focuses precisely on the issues glossed over by the expansive and pluralistic notions of "matter" and "nature" outlined above, including the challenge of tracing and evaluating fundamental underlying conflicts. In the posthuman preserve of a pluralist "nature" in which toxins rub shoulders with lifeforms, brush fires with log cabins, greenhouse gases with coral reefs, the attempt to abstract away from any kind of an overly "-centric" perspective poses challenges for concrete forms of judgment, politics, and critique.

Proponents of new materialism would surely challenge the notion that their approaches fall into this category. Bennett, for example, mobilizes reenchantment to unearth those moments that challenge the totalizing power of capital. However, the notion that capitalism thrives on the mythology of totalization may already be outmoded. In the face of the hyperactivity of the various devices, platforms, and applications that seek to act upon us—users—in myriad ways, the pressing task of the coming era of passive interactivity will be to excavate the imperatives, interests, and intentions built into the network—that is, the ways in which the responsive environments we inhabit are structured and shaped to reflect particular interests at the expense of others. It is not enough, in other words, to unfold the indefinite list of agentic participants in a particular assemblage (or, in a different register, all the possible contributing variables in the database). A critical approach must necessarily consider the fault lines that both permeate and disappear in such an analysis (this is the danger of operating at the level of assemblages)—and the interests served by this disappearance. A paradox is posed by attempting to enlist a sensitivity to matter's vibrancy—and the attendant openings and possibilities figured therein—in a system where intentionality as a political force has already been displaced, dissipated, and dispersed. The climate change debate, for example, reveals the capitalist right wing's version of "vibrant materialism": a strategic attunement to the myriad factors contributing to global warming and the debunking of an overweening anthropocentrism that imagines humans might be identified as either cause or cure. The enumeration of possible causes (Milankovitch cycles, ocean currents, fluctuations in solar radiation) aligns itself with the promulgation of uncertainty (there is no way we can really know—it's far too complex) and the dismissal of intentionality (even if we attempted to intervene, the ripple of our intentions would be drowned in a sea of unintended consequences and complex interactions). Critical analysis, in the face of such approaches, must consider the imperatives that inform the mobilization of uncertainty and the diminution of human agency (just as it must consider the imperatives that structure and are served by the data mining process, rather than simply noting the pattern of correlations that emerge). Surely complexity must be addressed, but in ways that are both amenable to discerning structuring conflicts and to enabling political intervention.

Three Dimensions of Interactivity

Increasingly, interactive devices and platforms are coming to serve as the locus for enhanced forms of participation in political and economic contexts. One of the results has been the tendency to conflate the two: Interactivity becomes a kind of participation-enabling capability. Thanks to distributed, digital, networked devices, it becomes easier than ever before to respond, share, organize, and distribute. To borrow from Jenkins, technological protocols bear upon social protocols. Simultaneously, an increasing reliance on digital platforms and devices redoubles participation in the form of interactivity: The apparatus captures data about our activities, our interactions, our participation. In this regard, interactivity is the shadow of participation, in the way that the digital glare of the Internet casts data shadows. The development of interactive devices and applications is shaped (although not uniquely determined) by the imperatives at work in various institutional settings—both commercial forms of influence and governmental forms of management, governance, and control. These imperatives are certainly subject to the vicissitudes of the material forms they take, to serendipity, surprise, and material resistance, but they are not eclipsed or subsumed by such forces. Indeed, the development of various forms of interactive capabilities follows the trajectory mapped out by these imperatives. If the online economy becomes increasingly reliant on data mining, then applications and devices are developed to enable more comprehensive data collection. Nic Carah has traced this evolution in the ways marketers incorporate branded opportunities into youth-oriented music festivals.[37] Early attempts required effort and some inconvenience on the part of users (connecting mobile phones at special stations to upload content), whereas more recent strategies are increasingly passive (wristbands that detect location and share information about who is following which musical acts). These strategies retain some traces of more active forms of interactivity (data is shared by particular individuals) but render it increasingly passive, like the Facebook app that listens to your music for you in order to share your listening habits more frequently and reliably.

With such examples in mind, we might differentiate between three dimensions of interactivity: that which comprises interpersonal interactions (between users and known or assumed contacts); that which involves intrapersonal interactions (between users and self-monitoring and quantifying devices that relay information back to the user); and that which refers to impersonal (or institutional/commercial) interactions (between users and the various entities that collect information about their activities and communications). Users tend to be much more aware of the former two dimensions, which are the intentional aspects of their online communications and interactions. They are generally less conscious of the various institutions and applications that are gathering data about them and of the ways in which their data is being used. Nonetheless, the imperatives of institutional data collection—especially in the commercial sector—shape the development of online applications and the services and infrastructures that support them. The emerging imperative is to foster frenetic, active participation in the realms of interpersonal and intrapersonal interaction, and passive (although comprehensive) interaction in the realms of impersonal or institutional interaction. Thus, for example, social media platforms are developed to encourage increasingly fast-paced forms of interaction between growing networks of users. If there might have been some self-consciousness on the part of users about having hundreds of "friends" on Facebook, this has fallen by the wayside in the drive to have as many Twitter "followers" as possible. Compared to the rapid-fire updates of text messages, tweets, and status updates, even blogs come to look somewhat ponderous, although they—along with other more substantial works—provide impetus for the forms of quick-hit self-promotion and linking enabled by social media platforms ("check out my recent blog post on *x*").

From the perspective of data collectors, frenetic interpersonal interaction generates increasingly high-resolution data profiles: We reveal ourselves in our interactions with others—selectively to particular audiences, but comprehensively to the unintended audience of the platform. Our Gmail correspondents are only aware of the messages we send to them, but Gmail itself sees all the e-mail we send to (and receive from) all our correspondents. The frenetic participation in interpersonal interactivity tends to mask the underlying audience. I have spent dozens of hours interviewing people about online privacy in the past two years, and one of the most persistent findings is that people tend to think about their privacy vis-à-vis their *intentional* audiences—that is, those with whom they are deliberately communicating, whether these be Facebook "friends," Twitter followers, or other correspondents and viewers of various kinds. They understand, for the most part, that they are crafting a particular type of portrayal for certain defined audiences, but rarely is the platform itself acknowledged as an intended audience, or even one that is given much consideration. Recent research has indicated that teens, for example, remain unconcerned about third-party data collection.[38] Tellingly, when people are asked to think about the platform in terms of such forms of data collection, the tendency is to personalize it—to consider what it might mean to have a particular person at Google or Facebook or Twitter or Snapchat read through one's messages or posts. And the result of personalization is to think about the potential consequences of comprehensive

monitoring in terms that are delimited by human capacities. Thus, typically, when the issue of privacy is raised, particularly with younger users
(in their twenties), a familiar response is to dismiss the notion that anyone
might be "interested" in the details of their lives, or to assume that their
individual information would simply get lost in the data deluge, whose
very size, like the urban environment before it, serves as a kind of guarantor of anonymity. Similarly, the potential uses of such data are viewed
through a human frame: Why would someone care if a complete stranger
whom one will never meet knows about the details of, say, one's personal
life; what could such a stranger possibly do with such information that
might have any impact on one's own life? We might describe this response
as a kind of humanization of the data gaze—an attempt to read it through
the lens of one's intentional interaction with others, as if the capacities of
one's friends and of Google were the same.

In actuality, thanks to automated forms of data storing, processing, and
retrieval, one's data does not get "lost in the crowd"—nor, of course, do
individual employees spend their time poring over it to extract interesting
information. As Google likes to put it, "no humans read your email."[39] But
such a response to concerns about data collection misses the point: The
paradigm of human attention, calculation, and comprehension falls short
when it comes to considering the potential impact and consequences of
data collection—as does the notion of perceived harm ("I'll start to worry
about all this data mining when I see some kind of negative impact on my
own life"). One of the characteristics of emerging forms of social sorting is
their opacity. If a company data mines job applications to determine who
might be a good employee, and determines that some candidates should
be ruled out because they used the wrong Internet browser when they
filled out their online job application, the applicants themselves are unlikely
to be able to make the connection between the data mining process and
the adverse impact on their lives.

In contrast to the attempt to read the affordances of impersonal interactivity through that of the interpersonal, what needs to be fostered is a
database imaginary adequate to the capabilities and uses of the machinic
"gaze." Such an imaginary would take into consideration the ways in which
the increasing scope of data collection is paralleled by the passive-ication
of data capture—that is, by the transition toward ambient data collection
built into the devices we use, the spaces through which we move, and the
digital networks upon which we increasingly rely for a growing range of

information and communication practices. The more data that is captured about us, the less participatory this process becomes—at least in the sense of active, conscious action that has some sense of and control over its consequences. Such an imaginary would take into consideration the link between increasingly active forms of interpersonal interaction, information search and retrieval, and the processes of passive-ication; it would trace the uses of data capture and mining that are not reducible to human-scale information processing.

The challenge faced by such an imaginary will be to encompass the unpredictability of the connection between categories of data and their potential uses. Addressing this challenge will, in turn, mean acknowledging the disarticulation of correlation and explanation. We are entering a world wherein the uses to which data can be put range far beyond our understanding of what counts as pertinent information. When I tell participants in focus group discussions that, for some job categories, data reveal that one's choice of Internet browser is a good indicator of future job performance, their initial response is, "But that information isn't relevant." For them, job-related data comprise hitherto established forms of pertinent information: past work experience, school transcripts, and so on. From the perspective of the data-driven decision-making process, relevance includes any and all data with predictive power, whether or not these data seem to have any intrinsic connection to the decision at hand. Moreover, no data can be ruled out in advance as irrelevant, thanks to the unpredictable (emergent) character of the data mining process itself; any aspect of one's life that can be tracked, collected, stored, and mined is, in theory, fair game. Only the mining process itself can determine its relevance.

Perhaps most importantly, any imaginary adequate to contemporary and future deployments of data mining practices will have to take into account the impact of relations of ownership and control on the opacity of decision-making processes. The critique of the political economy of big data will need to consider the ways in which situated imperatives inform the data mining processes. Otherwise the danger is that the algorithm will serve as a mechanism for laundering bias and masking power differentials. For example, potential employees do not recognize the adverse effect the data mining process has had upon them if they do not know why their application has been rejected. If they do not recognize this effect, they cannot respond to it. Much the same might be said of the proliferating forms of social sorting that are tapping into the logic of data mining: loan eligibility, insurance eligibility, educational opportunities, health-care decisions, and a growing range of assessments. There is a triangle of power at work whose dimensions need to be taken into consideration; at its apexes are those who have control over the decision-making processes that allocate resources, those who have control over the databases and infrastructures that are increasingly coming to inform those decisions, and those who control the malleable digital platforms that both enable ongoing experimentation and shape user awareness of the monitoring process and its outcomes. A platform like Facebook, for example, can run ongoing controlled experiments on its users because it controls the platform. By the same token, online platforms can implement forms of nontransparent social sorting—exposing consumers to different prices for the same product, for example, based on information that reveals some consumers are willing or able to pay more.

The implementation of a data-driven machinic gaze, in other words, confounds conventional expectations regarding how surveillance and information processing works. Since much of the conceptual and regulatory apparatus related to questions of privacy and forms of participation still operates within the scope of these expectations, it needs to be updated to reflect and anticipate emerging data mining practices. Absent such an update, we may find that active forms of participation online are redoubled by increasingly passive ones, amounting to automated participation in data-driven control systems.

Notes

1. Vince Horiuchi, "We Want Our MP3," *Salt Lake Tribune,* January 20, 2000, C1.

2. Jeff Howe, "Big Champagne Is Watching You," *Wired,* October 2003, http://archive.wired.com/wired/archive/11.10/fileshare_pr.html.

3. Kashmir Hill, "Facebook Wants to Listen in on What You're Doing," *Forbes,* May 22, 2014, http://www.forbes.com/sites/kashmirhill/2014/05/22/facebook-wants-to-listen-in-on-what-youre-doing/.

4. Mark Miller, "Shazam Bulks Up Ad Expertise with Mindshare Deal," *BrandChannel,* November 25, 2013, http://www.brandchannel.com/home/post/2013/11/25/Shazam-Mindshare-Ads-112513.aspx/.

5. Jamillah Knowles, "Republicans Have Less Diverse Music Taste Than Democrats," *TheNextWeb.Com,* July 12, 2012, http://thenextweb.com/insider/2012/07/12/republicans-have-less-diverse-music-taste-than-democrats-how-music-can-predict-our-political-leanings/.

6. Elizabeth Dwoskin, "Pandora Thinks It Knows if You Are a Republican," *The Wall Street Journal*, February 13, 2014, http://www.wsj.com/articles/SB10001 42405270230431500457938139356713007 8/.

7. Robert Lee Hotz, "The Really Smart Phone," *The Wall Street Journal*, April 23, 2011, http://www.wsj.com/articles/SB1000142405274870454760457626326166 79848814/.

8. Iain Thompson, "Microsoft's Moodscope App Predicts Smartphones' Users' Feelings," *The Register,* June 28, 2013, http://www.theregister.co.uk/2013/06 /28/microsoft_moodscope_phone_mood_detector/.

9. See, for example: Anonymous, "Robot Recruiters," *The Economist,* April 6, 2013, http://www.economist.com/news/business/21575820-how-software-helps -firms-hire-workers-more-efficiently-robot-recruiters/; also Malcolm Gladwell, "The Science of Shopping," *New Yorker,* November 4, 1996, http://gladwell.com /the-science-of-shopping/.

10. George Leopold, "'Social Physics' Harnesses Big Data to Predict Human Behavior," *Datanami,* May 21, 2014, http://www.datanami.com/2014/05/21/social -physics-harnesses-big-data-predict-human-behavior/.

11. This is doubly true when it comes to sharing information with Facebook. One report indicates that the company's instant messenger app even collects data about what orientation a phone is held in and how often. See Lauren O'Neil, "Facebook Messenger Found to Be Tracking 'A Lot More Data Than You Think,'" *CBC News,* September 12, 2014, http://www.cbc.ca/newsblogs/yourcommunity /2014/09/facebook-messenger-found-to-be-tracking-a-lot-more-data-than-you -think.html.

12. SoundCloud Help/Sharing Page, "How Do I Share to Facebook?" *Sound-Cloud,* http://help.soundcloud.com/.

13. "NameTag App," NameTag, http://www.nametag.ws/.

14. Ibid.

15. Susan Buck-Morss, *The Dialectics of Seeing: Walter Benjamin and the Arcades Project* (Cambridge, Mass.: MIT Press, 1991), 99.

16. Karl Marx, *Capital: A Critique of Political Economy,* Vol. 1 (Moscow: Progress Publishers, 1887), 80.

17. Slavoj Žižek, *How to Read Lacan* (London: Granta, 2006), 22.

18. Ibid.

19. Sara Ahmed rehearses some of the dimensions of this turn in "Open Forum Imaginary Prohibitions Some Preliminary Remarks on the Founding Gestures of the 'New Materialism,'" *European Journal of Women's Studies* 15, no. 1 (2008): 23–39.

20. Jane Bennett, *Vibrant Matter: A Political Ecology of Things* (Durham, N.C.: Duke University Press, 2009).

21. Jane Bennett, "Commodity Fetishism and Commodity Enchantment," *Theory & Event* 5, no. 1 (2001), doi: 10.1353/tae.2001.0006.

22. Ibid.

23. Ibid.

24. Ibid.

25. Ibid.

26. Ibid.

27. Slavoj Žižek, "Toward a Materialist Theory of Subjectivity," lecture at Birkbeck Institute for the Humanities, London, May 22, 2014, audio available online at http://backdoorbroadcasting.net/2014/05/slavoj-zizek-towards-a-materialist -theory-of-subjectivity/.

28. Alexander R. Galloway, "Poverty of Philosophy: Realism and Post-Fordism," *Critical Inquiry* 39, no. 2 (2013): 347.

29. Bennett, *Vibrant Matter.*

30. Ibid., 34.

31. Jeremy Packer, "Epistemology Not Ideology OR Why We Need New Germans," *Communication and Critical/Cultural Studies* 10, nos. 2–3 (2013): 298.

32. Galloway, "Poverty of Philosophy," 347.

33. Diana Coole, Samantha Frost, Jane Bennett, Pheng Cheah, Melissa A. Orlie, and Elizabeth Grosz, *New Materialisms: Ontology, Agency, and Politics* (Durham, N.C.: Duke University Press, 2010), 9.

34. Packer, "Epistemology Not Ideology," 298.

35. Marianne Jørgensen, "New Materialism and Political Potential: A Critique of Karen Barad's Agential Realism," paper presented at the 4S Annual Meeting, Copenhagen Business School, Frederiksberg, Denmark, October 17, 2012.

36. Iris van der Tuin and Rick Dolphijn, *New Materialism: Interviews & Cartographies* (Ann Arbor, Mich.: Open Humanities Press, 2012), 49.

37. Nic Carah, "Algorithmic Brands: A Decade of Brand Experiments with Mobile and Social Media," Public lecture, University of Queensland, June 12, 2014.

38. Mary Madden, Amanda Lenhart, Sandra Cortesi, Urs Gasser, Maeve Duggan, Aaron Smith, and Meredith Beaton, "Teens, Social Media, and Privacy," *Pew Internet & American Life Project,* May 21, 2013, http://www.pewinternet .org/2013/05/21/teens-social-media-and-privacy/.

39. Randall Stross, "When It Comes to Inbox Advertising, Less Is Still More," *New York Times,* May 14, 2011, http://www.nytimes.com/2011/05/15/business /15digi.html/.

CHAPTER 11

The Surveillance–Innovation Complex

The Irony of the Participatory Turn

Julie E. Cohen

O VER THE LAST SEVERAL DECADES, surveillance has become increasingly privatized and commercialized—and also increasingly participatory. This shift has challenged initial conceptualizations of surveillance as discipline and control by the state. Surveillance theorists have responded by turning to theories of networked flow, mass-mediated commodification, and performance to help explain the social, political, and psychological effects of surveillance. This chapter steps into that discussion, arguing that the effects of the participatory turn in surveillance are even more fundamental, and concern the extent to which we understand surveillance as itself subject to regulation. The rhetorics of participation and innovation that characterize the participatory turn work to position surveillance as an activity exempted from legal and social control.

Contemporary networked surveillance practices implicate multiple forms of participation, many of which are highly organized and strategic. Commercial surveillance environments use techniques of "gamification" to motivate user participation. Legal strategies for open access to data and open technical standards also have emerged as important drivers of the participatory turn in surveillance. Many information processing initiatives that rely on personal information are framed as open access projects, and seek to exploit and profit from the intellectual cachet that rhetorics of openness can confer. The ascendancy of such strategies coincides with a concerted effort to shift the tenor of legal and policy discourses about privacy and data processing. Participants in those discourses position privacy and innovation as opposites, and align data processing with the exercise of economic and expressive liberty.

The resulting model of surveillance is light, politically nimble, and relatively impervious to regulatory constraint. Commentators have long noted the existence of a surveillance–industrial complex: a symbiotic relationship between state surveillance and private-sector producers of surveillance technologies.[1] The emerging surveillance–innovation complex represents a new, politically opportunistic phase of this symbiosis, one that casts surveillance in an unambiguously progressive light and repositions it as a modality of democratic inclusion and economic growth. Within the surveillance–innovation complex, participation and commodification are entwined as a matter of political economy. But the surveillance–innovation complex is also a discursive and ideological formation. The rhetorics of participation and innovation advance the instrumental goal of holding the regulatory state at arm's length.

Playing and Being Played

Within commercial surveillance environments, the themes of play, games, and gaming are increasingly prominent. This participatory turn in surveillance challenges both legal understandings of power that rely on notions of discipline and theoretical constructions of power emerging from surveillance studies that rely on notions of control. To an increasing extent, commercial surveillance environments inculcate an orientation toward continual self-monitoring and the continual pursuit of feedback-based rewards. Networked selves in commercial surveillance environments are involved in a process of cultural reconfiguration that entails becoming gamers. We are playing, but we are also being played.

Consider first the example of Foursquare, a social networking application used to share information about one's whereabouts. Foursquare is generally credited with popularizing the idea of "gamification," the application of concepts and techniques derived from games to foster styles of "engagement" that promote business objectives in other areas of activity.[2] For the first four years of its existence, Foursquare offered subscribers opportunities to compete for rewards, which took the form of badges that might designate a subscriber "Mayor" of her favorite bar (for being a regular visitor) or "Player Please!" (for checking into the bar with three or more members of the opposite sex). Discount fashion retailer H&M, in partnership with an online gaming company, has used gamification to bring customers off the street and into its stores, offering players special items that can be scanned in H&M stores to generate discounts. Groupon, a social shopping site, uses gamification—in the form of an anthropomorphic icon named "Clicky" who entices users to pursue access to additional content and exclusive discounts—to motivate bargain hunters to visit the site more frequently. Nike+, a personal fitness tool, uses gamification to help its users set fitness goals, monitor their progress, and measure themselves against other users. In each of these examples, the gamified environment is a commercial surveillance environment: Personal information is collected from subscribers both during enrollment and throughout the course of play, and that information is used not only to deliver rewards but also—and more importantly—to conduct various forms of targeted marketing.

As the examples suggest, the gamification of commercial surveillance has a diverse set of origins. At its core, the gamification dynamic is a social one, based on motivating individual customers not only to play but also to recruit other players.[3] Gamification therefore may be understood, in part, as a strategic approach to commercializing the social. Foursquare emerged at a time when social networking platforms were migrating into the economic mainstream and seeking sources of capital financing. Its use of rewards as incentives for participants was both a strategy for achieving market penetration and a way of responding to potential investors' demands for a plausible revenue model. Foursquare is also a cautionary tale, however, because its gamification strategy proved incapable of holding subscriber interest, and in 2013 it announced that it was abandoning its badge system. A different and more durable example of gamification within a social networking platform is the unending competition for followers, favorites, and retweets on Twitter. While Foursquare's badge system proved, in the end, to be a passing fad, Twitter's metrics leverage and reinforce a more fundamental motivation for social influence more generally.

The gamification of commercial surveillance also has roots in direct marketing and customer loyalty programs that are decades old. Gamification, however, rewards customer loyalty in ways that generate public, social recognition. Again, the field of crowdsourced promotion has its own cautionary tales. Facebook's ill-fated Beacon service, which automatically coopted its members' social updates as promotional tools (and which was not gamified), resulted in widely publicized class-action litigation that culminated in a multimillion-dollar settlement. Contrast, however, the experience of crowdsourced promotion ventures that are shopping-oriented first and foremost: The USA Network's "Club Psych" website, which enables viewers of its television show *Psych* to compete for points redeemable for promotional merchandise, has significantly increased both merchandise sales and advertising revenues; entertainment giant Warner Brothers' "Insider Rewards" program rewards gameplay and social sharing with credits redeemable for additional content; shoemaker Jimmy Choo and car manufacturer Mini have staged wildly popular citywide scavenger hunts to promote their products; and so on.[4] Groupon's (momentary) success can be traced to its founders' recognition that customers could be made to absorb producer surplus and reveal information about their resources and patterns of discretionary spending at the same time. Gamification therefore may be understood, in part, as a set of techniques by which purveyors of goods and services make visible and leverage the social power of commercial networks.

Finally, the gamification of commercial surveillance environments also has roots in the Quantified Self (QS) movement, which was founded in 2007 by a group of Silicon Valley technology evangelists seeking better living through data. The initial impetus behind the QS movement was aggressively populist. QS entrepreneurs and communities offered participants the opportunity to shift control of health, diet, and fitness away from impersonal providers offering cookie-cutter recommendations and back toward individuals, and promised to maintain participants' data within walled gardens. Predictably, however, commercial providers of QS technologies and applications have now entered the field. As they have done so, the dialogue around QS has shifted, deemphasizing user control over data flows to third parties and emphasizing instead the need to provide and share data in order to gain new tools for controlling not only health, diet, and fitness, but also work habits, sex life, sleep patterns, and so on. Where the populist QS discourse was earnest and geeky, commercial QS products speak to lifestyle concerns in the language of marketing. At the same time, the gamification of QS environments makes both participation and sharing pleasurable.[5]

Gamified surveillance environments are not games, but they are like games in the sense that they manifest what Alexander Galloway calls diegetic actions (i.e., narrative-advancing actions framed by the narrator's first-person perspective) and nondiegetic actions (i.e., actions originating outside the world of gameplay that establish the frame and the conditions within which play may occur).[6] Gamified surveillance environments supply examples of each kind of action performed by both the operator (the human subject of surveillance) and the machine. Diegetic actions performed by the subject of gamified surveillance—actions that perform the in-world rituals of gameplay—include actions to unlock benefits or "level up" one's membership. Diegetic actions performed by the machine include the repetitive updates that create the background displays of the gamified surveillance environment, such as status updates from one's friends (Foursquare) or fitness community (Nike+) or ticker updates offering a continual stream of discount opportunities (Groupon). Nondiegetic operator acts include setup actions—for example, entering and saving preferences—and also the cheats and hacks that alter the flow of gameplay to the operator's benefit—for example, one widely publicized Foursquare cheat that enabled users to check into a location remotely by entering the target venue's identification code and GPS coordinates. Nondiegetic machine acts are the encoded actions that permit enrollment, determine the tiers of membership levels and corresponding benefits, and impose "gamic death" when a contest ends or upon logout.

Taking seriously the idea of a distinction between operator and machine acts, however, also exposes profound differences between games and gamified surveillance environments. The gameworld purports to be the social world, or some segment of it (social shopping for H&M and Jimmy Choo, fitness communities for Nike+, and so on). Instead of making the machinic rhythms of an imagined gameworld visible, the diegetic actions of a gamified surveillance environment promise to make social rhythms visible. Their focus is often quite narrow, however. Groupon and H&M, for example, used discount offers as experiments in surplus extraction, targeting and nudging patterns of discretionary spending. To similar effect, Foursquare's gamified maps highlighted patterns of leisure mobility. The nondiegetic acts of enrollment/login and logout/gamic death, meanwhile, have come to function as increasingly important gatekeeping features of networked existence. In gamified surveillance environments, the boundary between the gameworld and the real world within which it is embedded can be difficult to identify—and yet the gameworld is not the real world but rather a particular representation of it. Operator acts, meanwhile, are always already machinically mediated. Cheats and hacks, which function as acts of critique and intervention because they identify and call into question the gameworld's encoded rules, nonetheless often internalize the gameworld's overriding goals. As in immersive gaming environments, beating the system requires internalizing its implicit politics.[7]

Marketing strategists describe the relationship between gamified surveillance environments and games as imitative, but viewed in broader perspective it is more nearly parodic. By this I mean to do more than simply invoke Ian Bogost's characterization of gamification as "bullshit."[8] Although it seems correct, taxonomically speaking, to understand the vocabulary of gamification as puffery driven by marketing imperatives, gamification is not just messaging. Again the comparison to games as environments for mediated action is instructive. If, as Galloway puts it, the gameworld represents a kind of utopia that takes the form of an enacted allegory, then the gamified surveillance environment is an authoritarian allegory in which gamic action becomes both a commodity and a method of behavioral conditioning.[9] Raczkowski describes the way the "token economies" characteristic of gamified surveillance environments have been used as a form of behaviorist therapy for psychiatric patients, preparing them for reintegration into society by giving them sets of situation-specific rituals to perform.[10] In commercial surveillance environments, gamification takes on a similarly ameliorative gloss, inculcating repetitive behavior patterns oriented toward self-betterment.[11]

At this point it should be apparent that there is a kind of power at work in fully or partially gamified surveillance environments that neither legal scholars nor surveillance studies scholars have fully described. Most legal theorizing about surveillance and power has focused first and foremost on the state, and has framed power explicitly or implicitly as state monopoly of force. From a purely taxonomic perspective, that understanding is outdated. Foucault argued that the condition of modernity was characterized by the emergence of the capacity to discipline populations through organization and statistical normalization and by a shifting of pastoral power (i.e., the power to define the individual as cultural subject) from the church to the state. Both of these conceptions of power attenuate the connection between power and violence and align power instead with knowledge practices that discipline though habituation. According to Deleuze, the prevailing modality of power in the information age is no longer discipline, but rather control as manifested through the ability to direct flows of capital, information, and people.[12] Scholarship in surveillance studies has explored the ways that commercial surveillance activities enact

the themes of discipline-as-normalization and control-of-flows by capturing and directing flows of information to modulate individual behavior in
profit-generating ways.[13] Surveillance studies scholars also have explored
the intersections between surveillance and identity performance, examining the ways that exposure within surveillance environments affects both
identity and resistance.[14]

But the power at work in gamified systems of commercial surveillance
involves more than modulation of preferences and performances. Commercial surveillance environments are not simply disciplining and modu-
lating participation in the evolving global marketplace for goods and
services. They are also mobilizing participation in our own construction
as cultural subjects according to a very specific behavioral model. In the
context of commercial surveillance environments, the term "technologies
of the self" takes on a very literal meaning that refers to minutely quantified,
intensively monitored, feedback-driven trajectories of self-improvement.[15]
Commercial surveillance environments mobilize the pastoral power of digital technology, or the power of networked digital technologies to shape
narratives of the self, under economic conditions in which network users
are alienated from the process of shaping.[16]

The gamification of surveillance invests surveillance with a participation imperative directed toward production: a virtuous cycle in which
more is always more. As Jennifer Whitson describes it, "Becoming the
victorious subject of gamification is a never-ending leveling-up process,
guided by a teleology of constant and continual improvement, driven by
an unending stream of positive feedback and virtual rewards, and fuelled
by the notion that this process is playful."[17] The invocation of games and
play makes gamified surveillance feel natural and pleasurable, and this
helps to explain the ease with which gamified surveillance patterns migrate
from one context into others.

At the same time, and inevitably, the subjects of participatory surveillance are themselves reconfigured. The participatory turn in surveillance
denotes not just the genealogy of an activity or set of practices, but also
the genealogy of new forms of habitus that characterize digital natives, QS
mavens, bargain hunters, and so on. Operationally speaking, surveillance
alters the spatial and informational parameters that constrain and channel
evolving subjectivity, guiding individual action along more predictable
lines.[18] Gamification supplies this dynamic with a new legitimating narrative. Traditionally, answers to the question "What is surveillance for?"

have sounded in grim themes of necessity, security, and efficiency. Gamifi-cation offers a lighter and more appealing answer, investing commercial surveillance practices with an internal narrative arc and investing the gamer-self with socially reinforced conceptions of virtue. To be a player is the highest form of value.

Crowdsourcing and Crowds as Resources

Participatory models for commercial surveillance often are paired with legal and public relations strategies for open innovation. Such strategies may include contractual frameworks for open access to data and/or proto-cols and platforms, as well as public appeals designed to motivate the crowd-sourcing of surveillance projects. Among scholars who study the history and sociology of labor movements, there is now a rich debate about whether open development strategies that rely on crowdsourcing are democratiz-ing or exploitative. My aim here is not to take sides on that question, but rather to argue that crowdsourcing strategies for commercial surveillance have another, more fundamental purpose: to constitute a pool of informa-tion resources—the biopolitical public domain—that is available to be exploited by global information enterprises. Like the public domain in intellectual property law, the biopolitical public domain is not a specific data set but rather a legal construct: It supplies the background against which both the commercial entitlements and the knowledge production practices of the information age are being defined.

Both nationally and globally, many development projects that rely on personal information are framed as open access projects, and seek to exploit and profit from the aura of democratization that a formal designa-tion as "open" can create. In the public sector, health and education infor-mation in particular have become vehicles for open innovation. Beginning in 2010, the U.S. Department of Health and Human Services has convened an annual developers' forum now known as the "Health Datapalooza" conference, the theme of which is applications development for improved health-care delivery and medical records management. In 2012, the Depart-ment of Education jumped onto the bandwagon, hosting its first "Educa-tion Datapalooza" for applications developers. The website for the Obama administration's Open Government Initiative highlights strategies for open access and open innovation in these and other areas. In the private sector, the open hackathon has become both a recruiting tool and a method of generating research and development contributions.

Open development projects that rely on personal data engage in multiple kinds of crowdsourcing. First and most directly, such projects use open source software development models to harness the labor of applications developers. Second, the crowdsourcing of commercial surveillance also may mobilize individual participation in the labor-intensive processes of information collection and verification. Both types of crowdsourcing broaden cultural and technological participation but also reinforce capital's power to command labor. Developers and consumers gain a measure of agency but also double as voluntary information workers.[19] The idea of democratizing access to big data beckons seductively, but there are often significant gaps in the information provided to participants about project scope, about follow-on control over the information to be furnished by participants, and about the provenance of other information acquired by the system operators. Here it is useful to consider more carefully the motivations of the providers and financial backers of commercial surveillance environments. Why, for example, would Foursquare build a "map for nothing," and why would venture capitalists invest in it?[20] The answer is that what may appear to users as a map for nothing appeared to Foursquare and potential funders as a population map. There is a third, relatively undertheorized meaning of crowdsourcing, in which the crowd is a source of unremunerated labor but also of raw material to be used in processes of surplus extraction.

Discussion of informational capitalism often focuses on its outputs, but it is also important to focus on inputs: on the resources that a system of informational capitalism requires to function. In an earlier era, the transition from agrarianism to industrialism entailed the commodification of a set of important resources: land, labor, and money.[21] Since the early twentieth century, a second great transformation has been under way, from industrialism to informationalism.[22] The emergence of informational capitalism has entailed the commodification of other types of resources, including not only knowledge, cultural goods, and information processing capacity, but also attributes, preferences, and attention. The developing constellation of political and institutional practices relating to commercial surveillance gives concrete meaning to another bit of Foucauldian terminology: the idea of biopolitics. Contemporary commercial surveillance activities deploy the instrumentalities of informationalism to map, rationalize, and monetize populations.

Here the logic of global intellectual property markets provides a useful insight into the extension of commercial surveillance environments. Scholars who study the global intellectual property system have mapped a distinctive pattern of information flow, in which resources extracted from the global South flow north twice: once as indigenous resources extracted and appropriated by intellectual property industries headquartered in the global North and a second time as payments exacted for products based on those resources.[23] Within commercial surveillance environments, the world's populations are themselves the unexplored territory. Personal information processing has become the newest form of bioprospecting, as entities of all sizes compete to discover new patterns and extract their marketplace value. Practices relating to personal information processing constitute a new type of public domain, which I will call the "biopolitical public domain": a source of presumptively raw materials that are there for the taking and on which information-era innovators can build.[24] Often today, the most valuable information is that collected from consumers in developed countries, but the future of personal information processing is global. Particularly where health and genetic data are concerned, the gold rush has already begun. The push to constitute and exploit the biopolitical public domain is a contest over a postcolonial terrain, in which global networked elites seek to harness the power of populations worldwide.

The construction of the biopolitical public domain is bound up with a politics that is unapologetically and unironically a neoliberal biopolitics, oriented toward private appropriative freedom and inclined to regard protective regulation as invidious paternalism. So, for example, U.S. policy discussions about information privacy have long been framed in terms of the rational choice purportedly exercised by consumers in information privacy markets. Underneath the shiny patina of individualism that figures so prominently in policy debates, however, the mechanisms and outputs of commercial surveillance environments are not particularly individualistic. From the situated perspectives of the subjects of surveillance, the process may be experienced as personalized and immediately responsive to individual needs and wants. The marketing pitch for Fitbit, a QS fitness tool, states: "We believe you're more likely to reach your goals if you're having fun and feeling empowered along the way"; the public-facing website maintained by data broker Acxiom invites consumers to "help ensure you see offers on things that mean the most to you and your family."[25] From a systemic perspective, personalization within modulated information environments is pattern driven. Its point is to assimilate individual data profiles within larger patterns and nudge individual choices and preferences in directions that align with those patterns.

Against this background, rhetorics about inclusion and participation do important and unacknowledged normative and political work, compensating for a nearly complete lack of transparency regarding algorithms, outputs, and uses of personal information. The crowdsourcing framework also signals a subtle shift in the participation imperative, away from the libertarian vocabulary of individual choice and toward a more communal vision of inclusion in a common cause. Open data practices, we are told, do more than just promote lower prices, better service, and other forms of heightened personal utility; they also advance knowledge, improve lives, and promote the public good. Or, to put it differently, within gamified surveillance environments, "sharing is caring."[26]

The conception of the biopolitical public domain expressed by the emerging commercial surveillance economy, however, is a hierarchical conception. It reflects a biopolitics of crowds, and in particular a politics through which the "common productive flesh of the multitude has been formed into the global political body of capital."[27] The rhetorics of inclusion and participation that characterize commercial surveillance environments make the biopolitical public domain a particularly potent postcolonial formation, one to which the language of data protection regulation is an anemic and incomplete response. In an era in which the frontiers for biopolitical experimentation are continually expanding, regulating the new biopolitics of personal information processing effectively requires recognition of a commonality of interest that extends across traditional geographic, cultural, and socioeconomic fault lines. At the same time, the informational and material conditions of the participatory turn, and particularly its intensive personalization, make such recognition more difficult to imagine.

Returning, finally, to the open development practices with which this section began, it seems most useful to understand those practices as embedded within a complex politics. They are notionally democratic—anyone may become an applications developer or participate in a QS community—and yet in organizational and economic terms such projects often evolve and metamorphose in ways that partake of and reproduce the biopolitics of crowds. On such occasions, crowdsourcing and open access innovation feed the economic models they purport to resist, harvesting data for later commercialized use and providing a continual source of experiments about how best to gather, display, and harness its potential.

Participatory Governance and the Shape of Information Policy

Proprietors of commercial surveillance environments also rely on a third, complementary strategy for the participatory reconfiguration of surveillance, which involves the shaping of legal and policy discourses about privacy and the processing of personal information. In regulatory proceedings and in the popular press, the information processing industries have worked to position privacy and innovation as intractably opposed. That strategy has produced a discursive process that infuses "innovation" with a particular, contingent meaning linked to economic and expressive liberty. The framing of information processing as autonomous innovation signals an important shift in the political economy of surveillance: the emergence of a discursive formation that I will call the "surveillance–innovation complex." Within the surveillance–innovation complex, participation is framed as the result of uncoordinated, autonomous, inherently democratic choices made by free-market actors, even as participation and commodification are inextricably entwined. Advances in information processing are privileged as both innovation and expression, and regulatory oversight is systematically marginalized.

Over the past decade or so, in proceedings convened by the U.S. Federal Trade Commission, the Department of Commerce, and the White House, and also in the debates surrounding the proposed replacement of the European Data Protection Directive with a regulatory framework administered by the European Commission, members of the information processing industries have advanced a carefully crafted narrative organized around the themes of innovation and deregulation. Urging that "data-driven innovation can only occur if laws encourage use and reuse of data,"[28] they have argued that "industry self-regulation is flexible and can adapt to rapid changes in technology and consumer expectations, whereas legislation and government regulation can stifle innovation."[29] The rhetoric of freely flowing innovation as the lifeblood of the economy, and of regulation as its enemy, has been taken up by libertarian think tanks and technology blogs, whose contributors work to offset what they view as alarmist narratives about the extent of commercial surveillance.[30]

Regulators, for their part, have responded to the rhetoric of innovation by embracing the concept of a balance between two opposing goods. While rejecting the premise that regulation should take a back seat to innovation in the digital era, they have accepted the more general proposition that privacy and innovation are in tension. In a series of reports expressly framing the privacy–innovation relationship as one of conflict, they have argued that a predictable legal framework for privacy protection is necessary to create user trust and foster the right climate for market acceptance.[31] At the same time, however, American regulators and diplomats have worked to soften the European Commission's proposed new data protection regulation. Consistent with the interests of the powerful U.S. information industry, in U.S. interventions overseas the theme of innovation is uppermost.[32]

One way to understand the data processing industries' systematic deployment of rhetorics of innovation, openness, and autonomy is through the prism of rent-seeking. Pushback against threatened regulation or enforcement in the name of innovation and customer autonomy can work to deflect scrutiny of particular practices. But powerful actors do not simply deploy their considerable resources to garner one-time victories; they also use their resources to shape underlying legal and policy narratives about rights and obligations and to configure core legal institutions. In legal scholarship, the former dynamic has become known as "deep capture," while the latter is most closely identified with Marc Galanter's work on the ways that repeat players in litigation manufacture systematic, institutionally entrenched advantages.[33] Both dynamics shape the contemporary debates about information privacy regulation. Thus, for example, many scholars have noted that claims about consumer autonomy and consent work simultaneously to deflect regulatory scrutiny of commercial surveillance activities and to alter the prevailing sense of those activities as requiring supervision. And tech industry observers have long understood that strategic nonassertion of intellectual property rights in code represents a deliberate strategy for firms like Google, which invoke "openness" to bolster their dominance in markets for personal information. In terms of institutional structure, the data processing industry has generally acquiesced in the FTC's emerging practice of regulation by consent decree, and that stance is a calculated one. It accedes to cumbersome, but ultimately

procedural, audit and disclosure requirements in order to forestall regulatory imposition of more substantive standards of privacy protection.[34]

The rhetoric of information processing as innovation, and therefore
untouchable by regulators, represents a culmination of sorts for the deregulatory strategy. Implicit in that rhetoric is a conception of innovation as
an autonomous and inevitably beneficial process that is the natural result
of human liberty. Importantly, that conception is relatively invulnerable to
the standard science studies critique, for it does not depend on assumptions about the inevitable, linear nature of scientific and technical progress.
In the early twenty-first century, the idea of technological development
as autonomous has been thoroughly debunked. Instead, both industry
leaders and policymakers speak the language of diffusion studies, which
emphasizes all of the contingent factors that can affect the market uptake
of technological developments. The understanding of diffusion studies
that is current in business and regulatory circles, however, is a specifically
market-centered one, and it puts a different kind of autonomy in play, one
residing in the market itself. According to that understanding, invention
may be historically and technologically contingent, but innovation is not:
Innovations rise to the top of the pack as a result of the choices of self-
interested actors, catalyzing a continual process of social and economic
betterment. And if innovation is autonomous, then what is produced is
what should be produced. According to this logic, regulators can only get in
the way, and when they do we are all worse off, so they should not meddle.

The view of innovation as both inevitable and riskless is all the more
remarkable because it is an anomaly. In the domains of environmental
regulation and food and drug regulation, regulatory regimes have long
endorsed the precautionary principle, which dictates caution in the face of
as-yet-unknown and potentially significant risks. In those domains, moreover, the precautionary approach is widely recognized as creating incen-
tive effects of its own, encouraging research and development in areas
such as clean manufacturing and energy production, safe drug delivery,
and the like. In the United States, the precautionary principle traditionally
has not been seen as relevant to information and communications policy.
That view of the world owes a great deal to the seeming immateriality of
information and to the perception that—with limited exceptions in the
fields of health, education, and financial market regulation—the free flow
of information generally works to reduce risks rather than to create new
ones. As the evidence mounts about the risks inherent in large, unsecured

data reservoirs that function as "databases of ruin," the view that data-based innovation is riskless is increasingly difficult to maintain.[35] Yet the information industries and libertarian tech policy pundits have continued to advance a carefully crafted narrative that paints the precautionary approach as rigid, "Mother, may I?" policymaking.[36]

Here, moreover, a final component of the participatory strategy comes into play, which involves the claim that information processing activities qualify as speech protected under the First Amendment to the U.S. Constitution. Among tech industry lawyers and lobbyists, efforts have long been underway to paint information privacy regulation as an abridgment of their clients' freedom to participate fully in the marketplace of ideas. In *Sorrell v. IMS Health,* the U.S. Supreme Court appeared to endorse those efforts, ruling that constitutional protection for speech extended to an information processing program used to target pharmaceutical marketing to physicians.[37] While the Court has not yet squarely confronted the constitutionality of information privacy regulation, *Sorrell* represents a milestone for this particular deep capture strategy, because its reasoning links commercial information processing activities tightly to the exercise of fundamental liberties.[38]

The view of "innovation" that has emerged from these efforts is a complex and powerful construction, and its increasing dominance in policy debates fundamentally alters the political positioning of surveillance. Scholars of surveillance studies have mapped the contours of a surveillance–industrial complex in Western political economy: a symbiotic relationship between state surveillance and private-sector producers of surveillance technologies.[39] The surveillance–industrial complex encompasses a set of essential production relations: For surveillance technologies to be available, they must be produced, and for a market to exist, the technologies must be lawful and sought-after. Politically, however, the idea of a mutually beneficial relation between the state and producers of surveillance technologies has always been problematic, underscoring the degree to which systematic, focused observation of individual activities can threaten fundamental civil liberties. The surveillance–innovation complex is not only a set of production relations but also a discursive formation. Linking surveillance with innovative freedom alters the threat calculus that is implicit, and often explicit, in the habitual ways of thinking and talking about surveillance in the contexts of national security, law enforcement, and public administration. Within the surveillance–industrial complex, surveillance is a necessary evil; within the surveillance–innovation complex, it is a force for good.

As a discursive formation, the surveillance–innovation complex coheres with other deregulatory themes in the contemporary political landscape. Like so many other elements in contemporary political discourse, it aligns with a neoliberal philosophy of law and government, which "propos[es] that human well-being can best be advanced by the maximization of entrepreneurial freedoms within an institutional framework characterized by private property rights, individual liberty, unencumbered markets, and free trade."[40] The characterization of players in surveillance games as autonomous and consenting feeds into the neoliberal framing of regulatory choices. But rhetorics of open, autonomous innovation do more than proclaim a political philosophy. They produce and ratify a particular distribution of legal privilege and entitlement that reflects the distinctive governmentality of neoliberalism. The surveillance–innovation complex both constitutes the biopolitical public domain and describes who may appropriate its resources without fear of liability or obligation.

Conclusion

The participatory turn in surveillance has been characterized as a technologically enlightened form of self-emancipation, but that characterization is too simple (and often disingenuous). Participation comes in multiple forms and serves multiple purposes. It follows that the bare fact of participation ought not to be unthinkingly elevated above other values. A number of more basic questions suggest themselves: Who participates? For what purpose? And with what consequences? In the contemporary era of commercial surveillance, careful attention to the context and character of participation is essential.

Notes

Thanks to Lara Ballard, Darin Barney, Travis Breaux, Biella Coleman, Kate Crawford, Mireille Hildebrandt, Margot Kaminski, Frank Pasquale, Neil Richards, Jonathan Sterne, Jennifer Whitson, and participants in the 2014 Privacy Law Scholars Conference for their helpful comments on a draft version of this chapter, and to Alya Sulaiman for research assistance.

1. Ben Hayes, "The Surveillance-Industrial Complex," in *Routledge Handbook of Surveillance Studies,* ed. Kirstie Ball et al. (New York: Routledge, 2012), 167–75.

2. Rajat Paharia, *Loyalty 3.0: How Big Data and Gamification Are Revolutionizing Customer and Employee Engagement* (New York: McGraw Hill, 2013), 65; Gabe Zichermann and Joselin Linder, *The Gamification Revolution: How Leaders Leverage Game Mechanics to Crush the Competition* (New York: McGraw Hill, 2013), 6–7.

3. Paharia, *Loyalty 3.0*, 81–82.

4. Ibid., 98–114; Gabe Zichermann and Joselin Linder, *The Gamification Revolution*, 154–56, 163–64.

5. Jennifer Whitson, "Gaming the Quantified Self," *Surveillance & Society* 11 (2013): 170–71, http://library.queensu.ca/ojs/index.php/surveillance-and-society/article/view/gaming/. On video gaming motivation generally, see Andrew K. Przybylski, S. Scott Rigby, and Richard M. Ryan, "A Motivational Model of Video Game Engagement," *Review of General Psychology* 14 (2010): 155–58.

6. Alexander Galloway, *Gaming: Essays on Algorithmic Culture* (Minneapolis: University of Minnesota Press, 2006), 6–8.

7. Ibid., 91–92.

8. Ian Bogost, "Gamification Is Bullshit," *The Atlantic,* August 8, 2011, http://www.theatlantic.com/technology/archive/2011/08/gamification-is-bullshit/243338/.

9. Galloway, *Gaming,* 96–104. The business literature on gamification identifies behavior modification as an important goal. On design for behavioral conditioning in the related context of machine gambling, see Natasha Dow Schull, *Addiction by Design: Machine Gambling in Las Vegas* (Princeton, N.J.: Princeton University Press, 2012).

10. Felix Raczkowski, "It's All Fun and Games . . . : A History of Ideas Concerning Gamification," in *Proceedings of DiGRA 2013: DeFragging Game Studies* (Atlanta, Ga.: Digital Games Research Association, 2013), 4–5, http://www.digra.org/wp-content/uploads/digital-library/paper_344.pdf.

11. On gamification as behavioral conditioning through appeal to "intrinsic motivators," see Paharia, *Loyalty 3.0*, 27–38, 73–89. On design for behavioral conditioning in the related context of machine gambling, see Natasha Dow Schull, *Addiction by Design: Machine Gambling in Las Vegas* (Princeton, N.J.: Princeton University Press, 2012).

12. Gilles Deleuze, "Postscript on Control Societies," in *Negotiations 1972–1990,* trans. Martin Joughin (New York: Columbia University Press, 1995), 177–82.

13. See, for example, Mark Andrejevic, *iSpy: Surveillance and Power in the Interactive Era* (Lawrence: University Press of Kansas, 2007); and Greg Elmer, *Profiling Machines: Mapping the Personal Information Economy* (Cambridge, Mass.: MIT Press, 2004).

14. See, for example, Kirstie S. Ball, "Exposure: Exploring the Subject of Surveillance," *Information, Communication and Society* 12 (2009); and John E. McGrath, *Loving Big Brother: Performance, Privacy, and Surveillance Space* (New York: Routledge, 2004).

15. Michel Foucault, "Technologies of the Self," in *Ethics: Subjectivity and Truth,* ed. Paul Rabinow (New York: New Press, 1997), 223–52.

16. Katarina Giritli Nygren and Katarina L. Gidlund, "The Pastoral Power of Technology: Rethinking Alienation in Digital Culture," *TripleC* 10 (2012), http://www.triple-c.at/index.php/tripleC/article/view/388.

17. Whitson, "Gaming the Quantified Self," 169.

18. On the relation between surveillance, subjectivity, and privacy, see Julie E. Cohen, *Configuring the Networked Self: Law, Code, and the Play of Everyday Practice* (New Haven, Conn.: Yale University Press, 2012), 135–50.

19. For a useful discussion, see Tiziana Terranova, "Free Labor," in *Digital Labor: The Internet as Playground and Factory,* ed. Trebor Scholz (New York: Routledge, 2013), 33–57.

20. Leighton Evans, "How to Build a Map for Nothing: Immaterial Labor and Location-Based Social Networking," in *Unlike Us Reader: Social Media Monopolies and Their Alternatives,* ed. Geert Lovink and Miriam Rasch (Amsterdam: Institute of Network Cultures, 2013), 189–99.

21. Karl Polanyi, *The Great Transformation: The Political and Economic Origins of Our Time* (Boston: Beacon Press, 1957), 71–75.

22. James R. Beniger, *The Control Revolution: Technological and Economic Origins of the Information Society* (Cambridge, Mass.: Harvard University Press, 1986), 291–93; Manuel Castells, *The Rise of the Network Society* (New York: Wiley-Blackwell, 1996), 14–18; Dan Schiller, *How to Think about Information* (Urbana: University of Illinois Press, 2007), 20–35.

23. Anupam Chander and Madhavi Sunder, "The Romance of the Public Domain," *California Law Review* 92 (2004): 1341–43.

24. On the fallacy of raw data, see danah boyd and Kate Crawford, "Critical Questions for Big Data: Provocations for a Cultural, Technological, and Scholarly Phenomenon," *Information, Communication and Society* 15 (2012): 666–68.

25. See "The Fitbit Story," Fitbit, http://www.fitbit.com/story/; and "About the Data," About the Data, http://aboutthedata.com/.

26. Dave Eggers, *The Circle* (New York: Knopf, 2013), 311.

27. Michael Hardt and Antonio Negri, *Multitude: War and Democracy in the Age of Empire* (New York: Penguin, 2004), 189.

28. Daniel Castro, director, Center for Data Innovation, Letter to Nicole Wong, White House Office of Science and Technology Policy, March 31, 2014, http://www.itif.org/publications/public-policy-implications-big-data/.

29. Michael Zaneis, senior vice president and general counsel, Interactive Advertising Bureau, Testimony before the Subcommittee on Commerce, Manufacturing, and Trade of the House Committee on Energy and Commerce, Hearing on Balancing Privacy and Innovation: Does the President's Proposal Tip the Scale?

March 29, 2012, https://energycommerce.house.gov/hearing/balancing-privacy-and-innovation-does-presidents-proposal-tip-scale/.

30. See, for example, Larry Downes, "A Rational Response to the Privacy 'Crisis,'" *Cato Institute Policy Analysis* 716 (2013), http://www.cato.org/sites/cato.org/files/pubs/pdf/pa716.pdf/; and Berin Szoka and Adam Thierer, "Targeted Online Advertising: What's the Harm and Where Are We Heading?" *Progress on Point* 16 (2009), http://www.pff.org/issues-pubs/pops/2009/pop16.2targetonlinead.pdf/.

31. U.S. Department of Commerce, Internet Policy Task Force, "Commercial Data Privacy and Innovation in the Internet Economy: A Dynamic Policy Framework," NITA (2010), http://www.nita.doc.gov/files/ntia/publications/iptf_privacy_greenpaper_12162010.pdf/; U.S. Federal Trade Commission, "Protecting Consumer Privacy in an Era of Rapid Change," Federal Trade Commission (2012), http://www.ftc.gov/os/2012/03/120326privacyreport.pdf/; White House, "Consumer Data Privacy in a Networked World: A Framework for Protecting Privacy and Promoting Innovation in the Global Digital Economy," White House (February 2012), http://www.whitehouse.gov/sites/default/files/privacy-final.pdf/.

32. See, for example, William E. Kennard, U.S. ambassador to the European Union, "Remarks at Forum Europe's Third Annual European Data Protection and Privacy Conference," December 4, 2012, http://useu.usmission.gov/kennard_120412.html/; and Cameron F. Kerry, general counsel, U.S. Department of Commerce, "Remarks to the European Parliament," Interparliamentary Committee Meeting on the Reform of the EU Data Protection Framework: Building Trust in a Digital and Global World, October 10, 2012, http://photos.state.gov/libraries/usau/231771/PDFs/general-counsel-kerry-libe-submission-10-10-12.pdf/.

33. Jon Hanson and David Yosifon, "The Situation: An Introduction to the Situational Character, Critical Realism, Power Economics, and Deep Capture," *University of Pennsylvania Law Review* 152 (2003): 129, 202–30; Marc Galanter, "Why the 'Haves' Come Out Ahead: Speculations on the Limits of Legal Change," *Law & Society Review* 9 (1974): 95–160.

34. For a comprehensive discussion of the FTC's consent decree practice, see Daniel J. Solove and Woodrow Hartzog, "The FTC and the New Common Law of Privacy," *Columbia Law Review* 114 (2014): 583–676.

35. Paul Ohm, "Broken Promises of Privacy: Responding to the Surprising Failure of Anonymization," *UCLA Law Review* 57 (2010): 1748.

36. See, for example, Adam Thierer, "The Problem with Obama's 'Let's Be More Like Europe' Privacy Plan," *Forbes,* February 23, 2012, http://www.forbes.com/sites/adamthierer/2012/02/23/the-problem-with-obamas-lets-be-more-like-europe-privacy-plan/3/.

37. *Sorrell v. IMS Health, Inc.,* 131 S. Ct. 2653 (2011).

38. In an additional irony, following the revelations by NSA whistle-blower Edward Snowden regarding the extent of U.S. government surveillance of

telecommunications and Internet activities, many of the same companies endorsing the First Amendment strategy have rushed to cast themselves as defenders of their users' civil liberties, downplaying their own responsibility for creating and maintaining the data that the government collected.

39. Hayes, "The Surveillance-Industrial Complex."

40. David Harvey, "Neoliberalism as Creative Destruction," *Annals of the American Academy of Political Science* 610 (2007): 22.

· IV ·

Participation and Aisthesis

· CHAPTER 12 ·

Preparations for a Haunting

Notes toward an Indigenous Future Imaginary

Jason Edward Lewis

The future, in fiction, is a metaphor.

—Ursula K. LeGuin, *The Left Hand of Darkness*

Participatory Prelude

My long term collaborator, the artist Skawennati, is Iroquois, a Mohawk from Kahnawà:ke near Montreal. She opened our presentation at The Participatory Condition conference with a greeting based on the Ohenton Karihwatehkwen (Thanksgiving Address) of the Iroquois people:

> Today we gather together and remind ourselves that we, the people, have been given the responsibility to live in balance and harmony with one another and with all the living things. We think of all the people who are here today, and of all those who would have liked to be here but could not make it; we think of the people we have met, and the ones we have not met as yet. We remember that all of the people who live on every part of Mother Earth are all connected, related, and bound together in the same circle of life. So we bring our minds together as one as we extend our greetings, express our love, and give thanks, to all the people here and everywhere. Now our minds are one.[1]

The address acts as a framing ritual used at gatherings to clear participants' minds of distractions, allowing them to focus on the task at hand and align themselves toward a common goal. That goal may be successfully negotiating a treaty, or deciding on community membership. Regardless of the particulars of the event, the address is an attempt to establish a shared starting point from which all present can participate in crafting a common experience. I invite you, the reader, to share in the greeting. May our minds be as one.

In what follows I define and expand on the notion of the "future imaginary," and propose that Indigenous people should be concerned with actively populating it. I then provide several examples of how we, as Indigenous people, might structure that future imaginary to our own ends, and argue that, in order to do so, we need to actively participate in *building* technology as well as in using it.

The Future Imaginary

The concept of the future imaginary grew out of the work I conducted with Skawennati as part of a research network we codirect called Aboriginal Territories in Cyberspace (AbTeC). AbTeC is a group of artists, academics, activists, and technologists who investigate methods whereby Indigenous people can be full participants in the shaping of cyberspace — the archipelago of websites, video games, virtual worlds, data constructs, and other networked places that many of us inhabit on a daily basis.[2] These territories have been formed overwhelmingly in a Western intellectual tradition based on the binary logic of digital operations.[3] AbTeC's founding goal was to open up these territories to formation and occupation by individuals coming from an Indigenous context.

Early in AbTeC's history, Elizabeth LaPensée (née Dillon), one of AbTeC's research assistants, conducted a review of representations of Indigenous people in video games.[4] What she found confirmed our anecdotal experience: Indigenous characters were rarely featured in video games and when they were it was most often in a way that instantiated numerous negative stereotypes. This echoed research into Indigenous representation in film, television, and also comic books regarding the negative effects of such stereotypical portrayals.[5] Given that the science fiction-themed examples found in these media often articulate dreams of our future selves, it was a short step to next consider how they might also create damaging visions of our future.

Having grown up as fans of novels, movies, and comic books that featured science fiction themes, the members of the AbTeC research team knew firsthand how powerful these stories were in shaping our own ideas about the future. Whether considering the Prime Directive and transporter beams from *Star Trek*, the Force and hyperspace from *Star Wars*, Asimov's "Three Laws of Robotics," or even "soma" from *Brave New World*, our thinking about what could be possible and what might be probable five hundred years from now were profoundly shaped by these stories.[6] In our discussions it became clear that all of us shared an overlapping set of understandings and expectations about the far future, based on reading or viewing the same set of stories. We began to refer to these shared visions as our common future imaginary, that is, the ways in which we imagine the social configuration, political structure, and technological reality one, seven, or twenty generations hence. And we realized that an important ingredient in creating change in the rate of Indigenous participation in the cyberspace of the present was to actively imagine Indigenous people in the cyberspace of the future.

The question then became, How do we populate that future imaginary with Indigenous characters, stories, and worldviews?

Imaginaries and Their Implications

The myths of an "empty land," a "dying people" contaminated by European culture and the Indian as Nature (and all its corollaries) reflect and shape present-day attitudes towards First Nations people.

—Marcia Crosby, "Construction of the Imaginary Indian"

The term "social imaginary" has been invoked academically in varying configurations since the 1950s.[7] For the purposes of this essay, I rely on Charles Taylor's definition of social imaginary as "the way ordinary people 'imagine' their social surroundings . . . not expressed in theoretical terms, but . . . carried in images, stories, and legends."[8] Taylor deploys "social imaginary" as part of his critique of modernity, specifically to examine how the European and American worlds were able to develop key cultural structures such as self-governance, a capitalist economy, and the public sphere. The social imaginary defines what "makes possible common practices and a widely shared sense of legitimacy."[9] Taylor speculates that the ingredients that form the social imaginary often begin as theories held by a (usually elite) few, which then slowly seep into the general populace until, at some point, they simply become part of the general background understanding of how things work.

In AbTeC, we think of the future imaginary as a component of the social imaginary, one that crystallizes an informal and diffuse set of beliefs about how our culture will look one hundred, five hundred, or a thousand years from now. The North American future imaginary is one that all of us in AbTeC share, as participants in a more general North American social imaginary. It is what allows us to wonder, jokingly, what happened to the jetpacks we were promised, to ask one another whether we would prefer the red pill or the blue pill (*The Matrix*), to disgust one another with one-line references to chest bursters (*Aliens*), and to invoke instant anxiety about government hypersurveillance by tossing the term "pre-cog" into the conversation (*Minority Report*).[10]

These are playful engagements with a common social imaginary and its futuristic elements. Yet the contemporary social imaginary also contains creatures such as the "imaginary Indian." As described by art historian Marcia Crosby, the imaginary Indian developed out of the settler culture's need to erase the real Indian in order to depopulate North America in the minds of its immigrant population.[11] Anthropologists, collectors, and artists—believing they were "salvaging" the remnants of disappearing Indigenous cultures—created a narrative that served to reify a particular profile of the Indigenous individual and his community. This profile was then homogenized to encompass all Native nations of the Americas, to the point that it became the accepted definition of The Native.[12] Other narratives were excluded, including those given by Indigenous people themselves. This imaginary Indian, the "noble savage" whose culture of "primitive grandeur" had been irretrievably lost, became *all* Indians, shaping law, politics, social interactions, and cultural evolution for both settlers and Indigenous people from Contact to the present.[13]

In the process, the imaginary Indian became more real, and—it would seem from the general culture's embrace of him in preference to the reality—more sympathetic than any real Indian whom settlers might accidentally meet. He became *the* Indian of the social imaginary. "But," as the author Thomas King writes, "for those of us who are Indians, this disjunction between reality and imagination is akin to life and death. For to be seen as 'real,' for people to 'imagine' us as Indians, we must be 'authentic.'"[14] The legend of the imaginary Indian renders actual Indians mute and invisible. Taking our place is a prototypical Indigenous interlocutor who speaks only from a position firmly rooted in the past, on the margins of modern North American society, and of radically diminished agency. The imposition of this rhetoric of authenticity is one of the most pernicious consequences of the imaginary Indian.

The few representations of Indigenous characters that do appear in popular science fiction–themed movies and video games authenticate themselves against the imaginary Indian. Examples include the film *Avatar* (2009), in which the primitive aliens ride noble steeds into battle while wearing war paint, teach the white man how to honor mother Earth (or, in this case, mother Pandora), and ultimately turn to him as their only savior;[15] the novel *Snow Crash* (1992), in which the protagonist is chased by a sociopathic Aleut with "poor impulse control" tattooed to his head;[16] and the video game *Gun* (2005), in which Blackfeet and Apache characters are used as fodder for settlers pillaging the American West.[17] None of these characters are in charge; none of them are telling their own story; and if they are technologically adept, it is only as warriors. All of them are anachronistic.

Yet ready counternarratives can be found if creators would do the work to find them. In *Walking the Clouds: An Anthology of Indigenous Science Fiction,* scholar Grace Dillon illustrates how Indigenous authors have employed science fiction tropes to give shape to fully lived Indigenous lives. She uses their work to position science fiction as a "valid way to renew, recover and extend First Nations peoples' voices and traditions."[18] The stories in the anthology provide a much wider range of characters, settings, and motivations than those stories created out of a social imaginary infused with the imaginary Indian. They reconfigure authenticity in terms of lived continuity rather than in terms of a slavish adherence to a settler fantasy.

The paucity of Indigenous people in the future imaginary is troubling. The future imaginary influences how a culture thinks about its future. The settler culture's future rarely includes Indigenous people, and, when it does, it involves a lazy extrapolation of the imaginary Indian—an individual from a culture frozen in time three hundred years ago and whose main redeeming qualities are to provide a shorthand for the primitive, the natural, and the lost. If Indigenous people are not present in the future, one wonders why the settler culture need concern itself with what happens to us now. We will, after all, be gone soon enough.

We see AbTeC's current challenge, then, as populating the present social imaginary with fully empowered subjects of a future imaginary. That means creating Indigenous stories, epistemologies, and characters through which our peoples can articulate our dreams and aspirations, and make us present in the future.

Continuity and Presence

Aboriginal Territories in Cyberspace grew out of a series of projects produced by Skawennati and myself from the late 1990s to the mid-2000s. *CyberPowWow* (1997–2004) used what was then a cutting-edge software called The Palace, a graphical chatroom.[19] Today, in 2014, online chat is a regular part of many people's everyday routine. In 1997, though, it was almost magical to exchange real-time text messages with somebody anywhere in the world for free. The Palace not only allowed multiple people to chat simultaneously but also made those people visible through avatars that represented their presence in the virtual space (Figure 12.1). Skawennati seized on this tool as a way for Native artists to visibly participate in digital culture, and also as a way to connect communities separated by the vast distances of the Canadian landmass. *CyberPowWow* also represented a response to the question, "Can Native people make digital art?" Such questions came from both within the Indigenous community and outside of it, a reflection of how the imaginary Indian had shaped ideas about Indigeneity; engaging with contemporary technology was not commonly considered compatible with an "authentic" cultural practice.[20]

In 2002, we collaborated on *Thanksgiving Address: Greetings from the Technological World*, a playful extension of the traditional Iroquois Thanksgiving Address.[21] The traditional form continues on from the greeting that opens this paper, giving thanks to the earth, the water, the thunder, the medicine herbs, and much more. We felt it was time to add to that list a few lines from our technological world, for which we are also thankful. In a video triptych, we are seen eating a feast of turkey, yams, squash, stuffing, and cranberries. As we do so, we thank the Creator for the tools with which we survive and thrive in the modern world—tools such as the Internet, TCP/IP, and software for digital media manipulation. *Thanksgiving Address* illustrates how we might incorporate contemporary concerns into traditional forms of oratory in a relevant and respectful way.

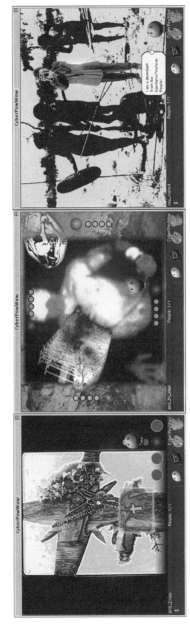

Figure 12.1. *Three CyberPowWow Palace rooms: R —Ahasiw Maskegon-Iskwew; M —Ahasiw Maskegon-Iskwew; L —Rea. CyberPowWow 2K, 2001.*

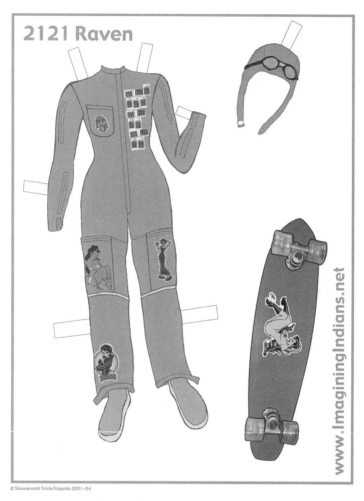

Figure 12.2. Imagining Indians in the 25th Century: 2121 Raven, 2001. Copyright
Skawennati.

In the 2000s, we started to more consciously focus on the future imaginary. This began in 2001 with Skawennati's *Imagining Indians in the 25th
Century,* a Web-based paper doll/time-travel journal. Commissioned for
the Edmonton Art Gallery's millennium exhibition, it is structured on a
timeline of one thousand years of Indigenous history. The timeline starts
two years before Contact, in the Tenochtitlán of 1490, and ends a thousand

years later, in 2490. Viewers navigate Katsitsahawi Capozzo, a young Mohawk woman, through ten points on the timeline, one per century, with each point highlighting an exceptional Indigenous event or person. At each stop Katsitsahawi dons an appropriate outfit and a journal entry captures her urban-Aboriginal perspective. For example, when visiting Chacotay, the Indigenous character from *Star Trek: The Next Generation*, she wears a Federation uniform and uses a tricorder (Figure 12.2).

Building on both the research and concepts from *Imagining Indians in the 25th Century*, Skawennati next created *TimeTraveller*™ (2008–2014).[22] A nine-episode machinima (cinema made in virtual worlds) series, *TimeTraveller*™ tells the story of Hunter (Figure 12.3), an angry young Mohawk man from the year 2121 who cannot find his way in an "over-mediated, hyper-consumerist North America" in which he has little connection to his Indigenous culture. Hunter employs *TimeTraveller*™, a technology akin to *Star Trek*'s Holodeck, but which uses special glasses as an interface.[23] The glasses allow users to "visit" historical events by immersing them in comprehensive interactive reconstructions. As in *Imagining Indians*, the main character explores places, people, and moments of importance to the Indigenous peoples of the Americas. Pre-Contact Tenochtitlán in 1490, the Dakota Sioux Uprising in 1862, the occupation of Alcatraz in 1969, the Oka Crisis in 1990 (Figure 12.4), and the "Manitouahbee Intergalactic Pow Wow of 2112" are some of the events Hunter visits.

Both of these stories connect the history of Indigenous cultures on this continent to Skawennati's vision for those cultures' possible futures. By stretching the arc of the series over such a long time frame, Skawennati creates a narrative of continuity and presence—one that stands in contrast and opposition to the story of disruption and disappearance that is often used to characterize the Indigenous history of Turtle Island. This continuity spans from a depiction of Tenochtitlán in 1490 to a future in 2121 where the Iroquois, the Cree, the Anishinabe, and the Blackfoot have all broken away from Canada to re-form independent confederacies. Both Katsitsahawi and Hunter are highly savvy about the use of advanced technology, and have integrated such technology into their everyday lives. Such continuity and characters serve to populate the Indigenous future imaginary with images of Indians who are able to draw upon their history in order to fully embrace a future where authenticity encompasses both the traditional and the contemporary.

Figure 12.3. Hunter Mega-figurine, 2012. *Copyright Skawennati.*

Figure 12.4. TimeTraveller™ Production Still: Face-Off, *2011. Copyright Skawennati.*

Phantasms and Hallucinations: Troubling the Monocultural Stack

"Thrones and dominions," the Finn said obscurely. "Yeah, there's things out there. Ghosts, voices. Why not? Oceans had mermaids, all that shit, and we had a sea of silicon, see? Sure, it's just a tailored hallucination we all agreed to have, cyberspace, but anybody who jacks in knows, fucking knows it's a whole universe. And every year it gets a little more crowded."

—William Gibson, *Count Zero*

Modern computing systems work via a very narrow logic, admit only certain kinds of information as data, and can perform operations representative of only a small, impoverished subset of the operations we enact as humans every day.[24] These systems exist as components of the "stack": the vertically interrelated and interdependent series of hardware configurations and software protocols that make high-level media computation and networking possible. The software stack sits on top of the hardware stack. Moving up the hardware stack is to move from circuits to microchips to computers to networks; moving up the software stack is to move from machine code to programming languages to protocols to systems. As you go upward, you are moving from custom solutions to generalized solutions, from specifics to abstractions. As you make this traversal from the deep structure to the surface interface, ever more of the details of the underlying configurations are hidden from you. With the increasing opacity, your ability to assert fine control over the execution of your algorithm decreases. Eventually you get to the software application or Web service layer of the stack. It is at this highly abstract level that most people interact with computational systems, as they use Microsoft Word, Snapchat image messaging, or Google search.

The technologists building the stack rarely acknowledge that bias infuses every layer of it, and with greater effect the farther one moves up and away from the physical constraints of the material substrate.[25] Software applications such as Word, for instance, can be thought of as orderly and (mostly) predictable assemblages of biases that reify the imagination of their creators into executable code.

Digital media scholar and computer scientist D. Fox Harrell proposes the term "phantasm" to describe a "particularly pervasive kind of imagination, one that encompasses cognitive phenomena, including sense of self, metaphor, social categorization, narrative, and poetic thinking."[26] He is concerned with how such phantasms become codified and reified within computer systems, how those systems perpetuate various forms of oppression, and how those systems could be built differently to better support "cultural content, diverse world views, and social values."[27]

Phantasms such as Crosby's imaginary Indian take up residence in a larger social imaginary that systematically misrepresents, delegitimizes, and dehumanizes the lived experience of contemporary Indigenous people. This social imaginary infects our own communities as well, where received concepts of authenticity and belonging are retained and applied to, for example, membership rules. An example is the use of "blood quantum" policies. Originally designed by settler-state governments to eliminate Indigenous peoples as distinct nations, band councils now use them to exile Indigenous people from their own communities.[28]

AbTeC's next phase, closely aligned with Harrell's efforts, is to address the damage caused by the phantasms that are instantiated in the basic structure of our hardware and software systems. Projects such as *TimeTraveller™, CyberPowWow, Imagining Indians in the 25th Century,* and *Thanksgiving Address* generate new kinds of imaginary Indians. Efforts such as the Skins Workshops, discussed below, aim to encourage the production of digital, networked media by Indigenous individuals. The next step is to engage with the production of the deeper technological layers that make such media possible. Our goal is for Indigenous people to become increasingly involved at all layers of the stack, to infuse it with alternative imaginaries. The step after that is for Indigenous people to participate in the design of alternative stacks, ones that are better able to accommodate Indigenous epistemologies, ontologies, and fields of action.

We are equally concerned with how these phantasms carry forward into the future imaginary of Indigenous and settler cultures. They restrict our participation in the creation of that future by confining us to the margins and denying our contemporaneity even as we are living it. They make it easier for settler culture to ignore those Indigenous future selves.

Bending Media

One of the central tensions in cyberculture is the one which exists between the command-and-control foundations of the underlying technological infrastructure and the expansive, emancipatory visions of some of its early pioneers. Douglas Engelbart imagined computer systems as a means to augment human intellect, Ted Nelson championed the notion of personal computers as "liberation machines," and Alan Kay sketched out how the technology could be used as an "intellectual amplifier" for education.[29] This tension has meant that cultural critiques of cyberculture have existed from its early days, though rarely as a central component of the discourse. In the last decade, though, it has become common to hear more discussion about how cyberculture continues to resist the participation of anybody other than the Western white male—at least in terms of who conceptualizes, designs, and implements the basic technology.

Harrell's analysis of phantasmal media engages the critical lineage within computer science that questions the epistemological assumptions that underlie both data structures and the algorithms that operate on them. Circuits, algorithms, and data are all culturally biased from the first moment. As a consequence, cyberspace has never been empty. It may have been—and still substantially remains—*terra incognita,* but it has never been *terra nullius.* It is haunted by the spirits of those who constructed it and also those who continue to modify the structures down deep in its stack. These individuals possess ideologies, dreams, and biases that find expression, in ways large and small, in how the plumbing was coded into being. They partook in the social imaginary of their time, one where the command-and-control vision of computing systems battled with the emancipatory vision. Both of these visions operated within a yet larger social imaginary. This imaginary constructed the logic that justified policies and practices of great detriment to the Indigenous people of the Americas, including second-class political status, residential schools, termination policies, land and resource theft, and so on.[30] Every layer of the stack springs from this same imaginary.

This is an old story. The history of Western media technologies' negative effects on Indigenous people post-Contact has been discussed in depth elsewhere.[31] The camera, for instance, was a technology we had not participated in developing, to which we had no access, and from which we received little or no benefit. Yet, in the hands of the European settlers, the technology shaped the way Indigenous people were seen, and thus treated, for centuries onward and thus played a key role in creating the imaginary Indian.

However, the story does evolve, even if slowly. Indigenous people now face a different reality than our ancestors, having regained some ground against the forces of eradication and assimilation. Indigenous creators working in photography, film, and video have documented their lives and the lives of their communities, bringing an endogenous perspective to the forefront and making it increasingly difficult for non-Indigenous scholars and creators to occupy the epistemological center of the discourse about Native life.[32]

Today, many Indigenous people have as much access as the settler population to the current new representational technologies. As with the later stages in the development of image- and time-based media, such access provides us with the opportunity to represent *ourselves* using the technology, in order to contest the representations made by the settlers. What is potentially more important, however, is the fact that cyberspace is still in its infancy, and thus still being shaped. Tech-savvy Natives can participate in that shaping: in conceptualizing, designing, and implementing the technology as well as in using it. We can learn programming, and hardware hacking, and how to construct new digital entities. Participating at this more fundamental level will greatly increase our ability to make the technology speak in the way we desire. We must always struggle against the deep structure on which cyberspace is built, but the nature of digital devices and networks is such that our ability to customize our cyber experience—to bend it toward our cultural context—and make the results available to the public is much greater than our ability to, for example, storm a network studio and take over the evening television broadcast.

Conceptualizing, Designing, and Implementing

When we began AbTeC we focused on claiming territory in cyberspace. We did this by creating and maintaining presence in cyberspace. We constructed scores of websites—to the point where our peers once jokingly asked us to leave some domain names for everybody else. We built a headquarters in Second Life called AbTeC Island to use as a base of operations and a virtual set for filming *TimeTraveller*™. We encouraged and supported other Indigenous people in producing digital, networked media that would represent their lives and communities. Our most prominent effort in this direction was the Skins Workshops on Aboriginal Storytelling and Digital Media Production (Figure 12.5).

The science fiction writer Neal Stephenson—when considering the ways that interfaces obscure the deep structures that make possible our everyday, unthinking use of computation—made a simple observation: "[I]f you don't like having choices made for you, you should start making your own."[33] The Skins Workshops are AbTeC's way of making our own choices by conducting grounded research into how best to integrate traditional stories with new media. We mounted five Skins Workshops between 2008 and 2013. One focused on teaching machinima and the other four on video game design.[34] Each video game workshop featured two hundred hours of instruction on subjects such as Aboriginal storytelling, image editing, sound design, 3D modeling, animation, and level design for digital games and virtual environments. Guest mentors, lecturers, and critics from major game companies (Ubisoft, Behaviour, and Minority) lent their time to bring the industrial perspective to the experience. We trained thirty (Mohawk, Anishinaabe, and Métis) participants between the ages of thirteen and forty from various reserves (Kahnawà:ke, Akwesasne, and Kanehsatà:ke) and urban communities (Montreal, Toronto, and Yellowknife).

The video games produced by the participants have won several international awards, and received a significant amount of press coverage

Figure 12.5. Skins 1.0, The Flying Head in the Village, from the Otsì:! Rise of the Kanien'kehá:ka Legends *game prototype, 2012. Copyright AbTeC.*

Figure 12.6. Ieniénte and the Peacemakers Wampum. *Game cover, 2014. Copyright AbTeC.*

(Figure 12.6). The attention has served to promote the idea of technologically savvy Indigenous people within our own communities and within the social imaginary of the settler culture. It has had the additional effect of proving to the participants how well and how powerfully they could shape the technology to speak with their own voice.

The next stage of the workshops will dive deeper down the stack. We are exploring how to incorporate more instruction in game programming into the video game workshops, as well as how to create workshops where programming skills are the focus. We are also exploring workshops on hardware hacking and DIY electronics, to provide participants with the ability to work at the hardware layer. We will continue to emphasize the importance of using Aboriginal culture as an inspiration for conceptualizing, designing, and implementing new media experiences while also highlighting how it is possible for Indigenous people to conceptualize, design, and implement the underlying technology. It is in so doing that Indigenous people will be full participants in shaping the future that awaits us.

Daring for Something Bolder

A metaphor for what?

—Ursula K. LeGuin, *The Left Hand of Darkness*

The writings of science fiction pioneer LeGuin capture the spirit of our idea of the future imaginary. Sketching visions of a future life—whether at the level of individuals, cultures, societies, or species—is rarely just about the jetpacks or the aliens or the minority reports. Jetpacks represent humanity's technical cleverness and desire to be unshackled from gravity; aliens represent the Others of our world, be they the primitive, the outcast, or the superhuman; and the minority reports represent living in a state of anxiety about the loss of free will in the face of a technologically overwhelming hegemonic state.

We are interested in building new sets of metaphors, new assemblages of phantasms—ones that will bridge between who we as Indigenous people are now and who we might be. By encouraging Indigenous people to become producers of digital, networked media, and eventually creators of digital technology, we hope to encourage our communities to embrace and engage what promises to be a highly technologized future. We need to build the capacity to participate in the construction of that future, and also validate, within our own communities, such activity as authentically Indian.

We must also be manifest in the future imaginary of the broader society. When our seventh-generation descendants explore whatever the latest incarnation of virtual space might be, they will find ghosts. Those ghosts will be the remnants of the epistemologies and ideologies that built those spaces—the phantasms in operation at their birth. Those ghosts will tenaciously inhabit cyberspace despite all attempts to exorcize them. We need to ensure that some of those ghosts will have been put there by Indigenous people who partook in the construction of the technological substrate, having been fully active in the future imaginaries that dreamed it into being. "We must dare for something bolder," says the curator and critic Paul Chaat Smith. "For those willing to leave behind the cheap, played-out clichés, a great project awaits."[35]

Notes

Conversations with my AbTeC co-director, Skawennati, have shaped my ideas on the Indigenous Future Imaginary from the first thought. I would like to thank Tamar Tembeck and Media@McGill for the invitation to speak at The Participatory Condition conference, and to those attendees who engaged our talk with their good minds. I am forever indebted to the talented artists, academics, and students at Obx Labs and AbTeC who have assisted me over the years as these ideas have developed. The research/creation effort underlying this chapter was supported by the Concordia University Research Chair in Computational Media and the Indigenous Future Imaginary; the Hexagram Research Centre; the Technoculture, Games, and Art Research Centre; the Social Sciences and Humanities Research Council; Le Fonds de recherche sur la société et la culture; the Canada Council for the Arts; and the Trudeau Foundation.

1. Jason Edward Lewis and Skawennati, "From CyberPowWow to AbTeC Island: Self-Determination in Digital Spaces," paper presented at The Participatory Condition conference for Media@McGill, at the Musée d'art contemporain de Montréal, Montréal, Quebec, November 15–16, 2013.

2. Jason Lewis and Skawennati Tricia Fragnito, "Aboriginal Territories in Cyberspace," *Cultural Survival Quarterly* 29, no. 2 (2005): 29–31.

3. Wendy Hui Kyong Chun, *Programmed Visions: Software and Memory* (Cambridge, Mass.: MIT Press, 2011), 138–78.

4. Beth A. Dillon, *Native Representations in Video Games,* video documentary, Aboriginal Territories in Cyberspace, Montréal, Quebec, 2011, https://vimeo.com /25991603/.

5. For research into film and television, see Sierra S. Adare, *Indian Stereotypes in TV Science Fiction: First Nations' Voices Speak Out* (Austin: University of Texas Press, 2005), 25; for research into comic books, see Michael A. Sheyahshe, *Native Americans in Comic Books: A Critical Study* (Jefferson, N.C.: McFarland & Company, 2008), 188–93.

6. *Star Trek (The Original Series),* television series created by Gene Roddenberry, NBC, 1966–69; *Star Wars IV: A New Hope,* film directed by George Lucas, 1977; Isaac Asimov, *I, Robot* reprint edition (New York: Spectra, 2008); Aldous Huxley, *Brave New World.* 2nd ed. (Toronto: Vintage Canada, 2007).

7. Cornelius Castoriadis, *The Imaginary Institution of Society,* trans. Kathleen Blamey (Cambridge, Mass.: MIT Press, 1998), 168–220.

8. Charles Taylor, *Modern Social Imaginaries* (Durham, N.C.: Duke University Press, 2003), 106.

9. Ibid.

10. *The Matrix,* film, directed by the Wachowski Brothers, 1999; *Alien,* film, directed by Ridley Scott, 1979; *Minority Report,* film, directed by Steven Spielberg, 2002.

11. Crosby, "Construction of the Imaginary Indian," in *Vancouver Anthology: The Institutional Politics of Art,* ed. Stan Douglas (Vancouver, B. C.: Talon Books, 1991), 267–90.

12. Audra Simpson, *Mohawk Interruptus: Political Life across the Borders of Settler States* (Durham, N.C.: Duke University Press, 2014), 96–114.

13. Paul Chaat Smith, "Home of the Brave," *C Magazine* 42 (Summer 1994): 39.

14. Thomas King, *The Truth about Stories: A Native Narrative* (Toronto: House of Anansi Press, 2003), 54.

15. *Avatar,* film, directed by James Cameron, 2009.

16. Neal Stephenson, *Snow Crash* (New York: Bantam Books, 1992).

17. Beth A. Dillon, *Native Representations in Video Games.*

18. Grace L. Dillon, *Walking the Clouds: An Anthology of Indigenous Science Fiction* (Tucson: University of Arizona Press, 2012), 1–2.

19. Skawennati Fragnito, *CyberPowWow,* http://www.cyberpowwow.net.

20. Loretta Todd, "Aboriginal Narratives in Cyberspace," in *Immersed in Technology: Art and Virtual Environments,* ed. Mary Anne Moser (Cambridge, Mass.: MIT Press, 1996), 179–94.

21. Jason E. Lewis and Skawennati, *Thanksgiving Address,* flash movie, http://www.obxlabs.net/shows/thanksgivingaddress/.

22. Skawennati Fragnito, *Imaging Indians in the 25th Century,* http://www.imaginingindians.net/.

23. D. C. Fontana and Gene Roddenberry, "Encounter at Farpoint," *Star Trek: The Next Generation* television episode, Paramount Studios, September 28, 1987.

24. Douglas R. Hofstadter, *Gödel, Escher, Bach: An Eternal Golden Braid* (Harmondsworth: Penguin Books, 1979), 285–309.

25. Terry Winograd and Fernando Flores, *Understanding Computers and Cognition: A New Foundation for Design* (Reading, Mass.: Addison-Wesley Professional, 1987), 70–79.

26. D. Fox Harrell, *Phantasmal Media: An Approach to Imagination, Computation, and Expression* (Cambridge, Mass.: MIT Press, 2013), ix.

27. Ibid., xv.

28. Pamela D. Palmater, *Beyond Blood: Rethinking Indigenous Identity* (Saskatoon: Purich Publishing Ltd., 2011), 28–54.

29. Douglas Engelbart, "A Research Center for Augmenting the Human Intellect," paper presented at the fall Joint Computer Conference, San Francisco, December 9, 1968; Theodor H. Nelson, *Computer Lib/Dream Machines* (self-published, 1974); Howard Rheingold, *Tools for Thought: The History and Future of Mind-Expanding Technology* (Cambridge, Mass.: MIT Press, 2000).

30. Vine Deloria Jr., *Custer Died for Your Sins: An Indian Manifesto* (Norman: University of Oklahoma Press, 1988).

31. Angela Aleiss, *Making the White Man's Indian: Native Americans and Hollywood Movies* (Westport, Conn.: Praeger, 2005); Beverly R. Singer, *Wiping the War Paint off the Lens: Native American Film and Video* (Minneapolis: University of Minnesota Press, 2001), 14–22; Edward Buscombe, *"Injuns!" Native Americans in the Movies* (London: Reaktion Books, 2006).

32. Vine Deloria Jr., *We Talk, You Listen: New Tribes, New Turf* (Lincoln, Neb.: Bison Books, 2007), 33–44.

33. Neal Stephenson, *In the Beginning . . . Was the Command Line* (New York: Avon, 1999), 151.

34. Beth Aileen Lameman and Jason Edward Lewis, "Skins: Designing Games with First Nations Youth," *Journal of Game Design and Development Education* 1, no. 1 (2011): 63–75.

35. Smith, "Home of the Brave," 42.

CHAPTER 13

Participatory Situations

The Dialogical Art of *Instant Narrative* by Dora García

Rudolf Frieling

Who is sure how to begin, begin anything? We put on ears, the ears of others, voices slip in, while the feet make rhythm, rhythm. She commands her partner to listen. There is a way in which what is embraced commands the body. A vibration, a pulses. One can only intake so much. When the feet can no longer response, an urge to leave, an urge to move on. We embrace an alien childhood. Not alien in a stupid sense. Just different. The shoulders drift forward. Ah, art. What makes this art. What is a portrait in another direction. Alarm bell is shrill. Is the art being attacked?

—Dora García, *Instant Narrative*

IN MEDIA ART, there is characteristically no "art" in the strict sense advanced by museum policy. There is no unique object that can be identified as an original valuable object. Instead there are only concepts, codes, instructions, copies, and relationships.[1] This condition applies to contemporary art in general, including the live arts, whose rise within the visual arts has been much discussed over the last decade. Where is the art when the art lies in the subjective experience of an enacted performance? Whether this performance is based on actors or software does not make a fundamental difference. The situations created by Tino Sehgal and collected by museums, including the San Francisco Museum of Modern Art (SFMOMA), make things even more challenging by provoking the question of where the art is when the museum does not even own a

Figure 13.1. Dora García, Instant Narrative, *2006–8 (installation view, SFMOMA, 2012); performance with software and video projection, dimensions variable. Collection of SFMOMA, Accessions Committee Fund purchase; photo by Johnna Arnold; copyright Dora García.*

certificate of authenticity, a set of instructions, or a file as verbally stipulated by the artist.

Writing an essay that addresses participatory or dialogical art at this moment in time, one gains the advantage of being able to directly shape an emerging critical and historical account. In the past, participatory works of art had long been considered anathema within the institutional realm of the museum. They were kept at bay by various mechanisms of exclusion and institutional expertise that called for a controlled environment inside of the museum. The rarefied museum object, crafted and signed by a recognized artist, was by definition the opposite of the open participatory work. Artists like Robert Smithson and Allan Kaprow were among the first to openly reject the museum, albeit for different reasons. They argued for an artistic practice outside the museum—then focused entirely on collectible objects, to the exclusion of site-specific interventions, such as those engaged by Smithson, or lived experiences, as in Kaprow's happenings.[2] From performative works to institutional critique, artistic practices

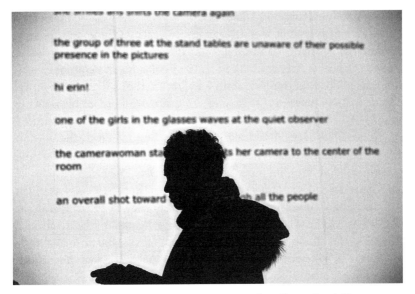

Figure 13.2. Dora García, Instant Narrative, *2006–2008 (installation detail, University of Michigan Museum of Art, 2014); performance with software and video projection, dimensions variable. Collection of SFMOMA, Accessions Committee Fund purchase; photo by Leisa Thompson; copyright Dora García.*

were among the first to challenge the notion of collecting finite works as commodities. The open and participatory work of art ran counter to the institutional framework of the museum. In other words, it tried to eschew the pitfalls of "museumification" and celebrate its uncollectible nature. The new critical approach that has emerged over the last decade in response to a surge of interest in open systems, collaborative practices, and participatory proposals elaborates participation in art as a "democratic" pursuit, but then criticizes this pursuit for its lack of ambition outside the realm of art.[3] Claire Bishop claims that participatory art "rapidly becomes a highly ideologized convention in its own right, one by which the viewer in turn is manipulated in order to complete the work 'correctly.'"[4] She goes on to call this phenomenon "pseudo-participation," that is, a kind of participation that simply conditions the public.

It is about time that we look at the specifics of what happens in participatory artworks in a given context and under different conditions. Rather than follow Grant Kester or Claire Bishop in retracing well-trodden paths that engage with specific works on an ideological meta level, I propose to examine the aesthetic, contextual, and experiential factors of a dialogue between a specific artistic situation and its public in a museum or exhibition space. The situation in question is Dora García's *Instant Narrative,* a work that has been ongoing since its inception in 2006. Introducing notions of dialogue and process into museum exhibitions has constituted a paradigm shift over the last several decades, but one needs to question the specific dialogue that is being staged or enacted. It can remain a simple gesture or be void of any new experience. In fact, historic examples of closed-circuit and real-time communication art projects such as Allan Kaprow's *Hello* (1969) simply confirmed that a dialogue was happening at all.[5] Artists have moved on from this early investigation of the mechanics of systems and distribution channels and now embrace the various semantics of who is talking, to whom, and about what. I first asked these questions in the context of an exhibition I curated at SFMOMA in 2008, "The Art of Participation: 1950 to Now."[6] In this essay, I am interested in exploring a new question—a question that one must raise in this period when participatory art is surging *within* institutions: Has the institution of the museum changed or has participatory art changed? It is to our advantage that we can begin to look back at the first histories of participatory art within the museum. Furthering these historical perspectives is one of the objectives of this essay. What follows is a close reading of one particular situation, an artwork that only emerges through dialogue and establishes a presence in the galleries of the museum when it is performed. In this situation, participants can identify effects, actors, interpreters, or performers, and experience their status as active agents of the artwork. As this essay sets out to show, this dialogic dimension indicates that the museum and participatory art have both changed interdependently: in relation to each other, in juxtaposition to each other, and, finally, in a continuous dialogue with their participants.

The institutional "locations" of performative and participatory art have multiplied, going beyond the spaces dedicated to public events such as theaters or lobbies to include the gallery spaces that have historically been reserved for the contemplative display of artworks. While the proliferation of time-based media in galleries has similarly blurred boundaries over the last two decades, effectively blending cinema and museum and gallery spaces under the umbrella of the black box, the experiential factor of

Figure 13.3. Dora Garcia, Instant Narrative, 2006–8 (installation view, Venice Biennial, 2011, as part of the performance installation The Inadequate); performance with software and video projection, dimensions variable. Collection of SFMOMA, Accessions Committee Fund purchase; photo by Rudolf Frieling; copyright Dora García.

participatory art is enhanced by how it stages an open situation that is potentially beyond the control of any institutional framework. Works like García's *Inserts in Real Time* address precisely this institutional condition, opening it up to the poetics of participation. My claim is not to deny the political readings of institutional critique but to foreground a dialogical relationship to the history, context, and experience of art. While a more comprehensive art historical analysis of these shifts within museum institutions is still to come, this closer examination of one specific participatory performance aims to contribute to a better understanding of these aesthetic and contextual dimensions.

Questioning and addressing the relationship between artist/performer and audience—the specific traditions and constraints of spectatorship—has a long tradition that one might even trace back to Dada, Bertolt Brecht, and other avant-garde events of the twentieth century. Some groundbreaking artists of the 1960s merit special mention for introducing the specific concerns of crossing boundaries between genres. There is a specific conjunction of artistic practices that have repositioned dance as a collaborative performance of everyday movements (Yvonne Rainer), reformulated sculpture as a series of changing material configurations (Robert Morris), and opened up a performative and participatory architectural space in closed-circuit installations (Dan Graham) with their live feed of real-time or delayed video images as a way to explore an electronic studio.

Today, artists interested in reviewing notions of change, collaboration, and participation do not necessarily place their practices outside the confines of the museum. They clearly operate under different institutional circumstances but often with a similar set of questions. What are the visitors when they are not visitors anymore but instead the subject of display and, in the case of Dora García's *Instant Narrative,* also characters in a narrative? What or who are readers when they become producers, and then perform in response to their reading? García, who also formulates this question in the work *WHERE DO CHARACTERS GO WHEN THE STORY IS OVER?* (2009), clearly believes that an artistic question has no answer. Her related series of golden-lettered wall sentences sums it up: *Una buena pregunta debe evitar a toda costa una respuesta* [A good question should avoid an answer at all costs] (2002). We will see how this is an apt introduction to a work that initiates dialogue between audience and place. Could we then deduce that a good dialogue avoids closure at all costs?

> six people in gallery, no one looks at the screen
> . . . Tall man in black jacket holds hands with a woman in a bright orange sweater. They are holding hands as they look at bulletin boards in the back gallery
> The look slowly
> observe
> The blonde girl, torquoise hoody, is back listening to the girl in orange.
> It must be an interesting piece.
> Many hoodies. Political moment.
> The other blonde girl—mom points to her presence on screen
> Fame is an interesting thing

Ever since Luigi Pirandello had six characters search for an author on stage, the self-conscious act of showing and telling has been a continuous presence on the larger "stage" of modernity. In a more political vein, Brecht's didactic dramatic plays have helped tear down the fourth wall of illusion, a theatrical tradition that is being reviewed by a contemporary generation of cultural agents. García is one of them. In her *Beggar's Opera,* she gave one of the minor characters of Brecht's *Threepenny Opera,* the apprentice beggar Filch, center stage at Münster's *Sculpture Projects* in 2007.[7] An actor, identified only by the number "6" on the plate he was wearing—a simple reference to the numbered list of works featured in both a map of the broader group exhibition and its brochure—roamed the streets of the West German city without a prescripted timetable. All of his interactions with Münster's citizens thus occurred without appeal to an explicit "art audience": those individuals roaming the streets of the city in search of various designated art locations. Any encounter between the actor and the audience was thus coincidental, and produced a series of aftereffects embedded into diary entries, newspaper reports, and, in one case, even a police report. For García's art audience, the "art" always seemed to be happening elsewhere.

García's hybrid practice involving art, literature, theater, and television has produced a body of work that is unique and prolific. She is an avid reader and huge fan of James Joyce—an obsession I share—and her references and direct artistic influences include Abbie Hoffman's 1970 *Steal This Book,* Lenny Bruce's legendary stand-up comedy, Erving Goffman's *The Presentation of Self in Everyday Life,* Franco Vaccari's pioneering work in participatory art, and Franco Basaglia's seminal work in antipsychiatry, to name just a few.

These criss-crossing interests converge in her focus on the politics and ambiguities of the place of art, taking the form of durational actions on real or virtual stages, often performed by actors who interrupt, question, or upset existing contexts. An early example is her project *The Kingdom,* "a novel for a museum" (first shown at MACBA, Barcelona, in 2003), in which the reality of all events in and around the museum was seemingly verified by their insertion into an ongoing online diary.[8] The diary was a record of events that had been facilitated by the institution: "The Kingdom's basic purpose is to disinform. The Kingdom relinquishes the spectator's education in favour of his/her perplexity. The Kingdom does not want to enrich the spectator's perception of the world, or his/her perception of him/herself: It questions them."[9] Asking the museum to be complicit in a strategy of disinformation is in fact asking a lot, going precisely against the inherent mission statement of all museums to serve and educate their public.

Conflicts, paradigms, and conventions of art and art audiences are at the core of García's work. Setting herself apart from the charged traditions of 1970s performances, her "inserts in real time" have shaped her hybrid performative approach for more than a decade. At the Venice Biennale in 2011, her four-month-long performance project and exhibition *The Inadequate* provided a big platform in the central gallery that was used by various performers and guest speakers, while the fringe of the platform was occupied by an inconspicuous performer continuously writing about what was going on in the spaces around her. The performer's writing was then projected into an adjacent gallery in real time. *Instant Narrative,* one of her key works, adopts a similar strategy. It was acquired by SFMOMA in 2011 and first installed there as part of the exhibition *Descriptive Acts* in 2012.

> The narrator arrives by the skin of her teeth.
> There is a hum of voices from the next room.
> A woman in ivory top enters room, puts on head phones and reads a
>   pamphlet.

"The narrator arrives by the skin of her teeth." Clearly, every text has a beginning, and this is one self-conscious yet poetic way of starting an "instant narrative" in the museum gallery before it opens to the public. For the first visitor who enters the gallery, however, this beginning might already be lost, since the production of constantly updated text will have already begun—just as paintings are always already on the gallery's walls—by the time she arrives. The performance thus rarely has a beginning that is visible to the first visitor, and its two components, a writer at a desk and a projected text, are not immediately recognizable as being part of a single work.[10] This apparent disconnect is intentional. The visitor sees the text and possibly the writer, but it dawns on the unsuspecting reader only gradually—or not at all—that the projected text is being produced in real time and that she is the one at the center of this writing performance. *Instant Narrative*'s text may be based on factual observation and yet include hypothetical deductions and musings that produce a sense of fictitious reality. In fact, there is no unmediated and unfiltered observation taking place. The text on the wall is a fragment of a prism, elucidating as much as obscuring the events in the gallery.

Two children watch the projection.
They talk amongst themselves.
A boy swipes at a girl, she dodges his blow.

A boy sits on a bench.
A girl sits on a nice modern chair.
as their mother calls they leave.

The museum gallery is first and foremost a setting for art, but there is an alternative narrative that is more basic: The gallery, like any other public space, is a charged sphere of social interaction. *Instant Narrative* was first performed in 2006 with the specific objective of observing people in an art setting, not through a surveillance camera but through the medium of a writer who is constantly at work but physically at a distance from the projected writing on the wall—possibly even in an adjacent room from which the gallery with the projection can be observed.

the two who were listening intently and yet less intently are taking
    pictures of their own intensity via the medium of the words.
the intense one says, "that one right there is about me"
and laughs.
and the other intense one who is less intense says, thank you
acknowledging that the power of seeing, describing, observing, is
    actually a kind of gift.

This observation—not via a surveillance camera, but via a writer who could nonetheless easily be mistaken for a staff member—generates a cybernetic feedback situation. It is literally and openly an experience of being talked about directly or referenced obliquely, but through mechanisms of feedback that are complex and delayed; *Instant Narrative* mediates experiences where the art talks about who we are, and not vice versa. An artwork, which is typically the object that is being looked at, turns the gaze around and looks back. As a matter of fact, it not only looks back, it also acts and talks in response—"it" being a hybrid of performer/writer and the white projection screen that displays a continuous narrative. Gallery visitors become "characters" in a narrative—or not, as they are sometimes neglected in favor of other events happening at the same time. Visitors cannot be sure that they will be "seen." Some start to act, others flee the scene and lurk outside the vision of the performer, and then there are those who reflect and respond in writing: "I want to disrupt the writer's control of the room. As people flow in and out of the space, I remain stubbornly present and still, continuing to scribble things down into my notebook. If I stand still, she can't write about me. Or perhaps this will give her license to look even *closer* at me."[11] After all, looking closer is a typical feature of television's instant replay. Here, however, looking closer might not reveal more detail but rather push the characters to act differently.

> wondering, still, she rifles through her belongs, smiling, wonder-
> ing. an onslaught of messages ready for comprehension, they
> both think, they both speak, they're both confused bored with
> their inertia they trace the walls one at a time, in their effort at
> forward movement, but even with their desired target found they
> concentration still wanders back along that path now seated in the
> dark, calling her thoughts back to her, she finds herself growing
> warmer and warmer, her outer layer now useless she removes it as
> it's burden falls away, she is only left more vulnerable. her deep
> breaths pass across the space, apparent to all, her voice thick with
> the season. sat against that furthest dull grey wall, she feels move-
> ment everywhere, footsteps and voices heard from afar, persistent
> beats transmitted through walls promise an elsewhere in it's
> freshest guise.

An ally of the restrained Samuel Beckett text rather than of surrealist automatic writing with its associative, creative expression, García's insistence that the textual process be based on factual observation produces a body of collective writing in which style becomes a function of the gallery's events or nonevents. Imagine: One would spell out loud all those observations that occur in one's head while walking through busy galleries. But what happens in a gallery when one is alone and no one else is around? What emerges when a Cagean silence is not only listened to but actually being described?

> the museum is winding down, winding down
> the theater voice drones on, fewer footsteps, more walkie-talkie
>   voices:

fewer visitors; more guards
the museum is closing soon
still echoes of voices
adrina the guard is back
she is looking foward to getting off her feet
the observer is tired and hungry
describing reality is hard work
exhausting, really
the observer can't believe she has to go see her mom after this
closing time!!!!!!

The artist had specifically instructed the writers not to slip into the first person singular, to sustain a distance from the events being described. Obviously, this is a loose framework; individual styles take over. The writing is at times personal and the authorial positions vary greatly, so do they all constitute the same artwork?

Two women enter the room
Mother and daughter? Student and friend?
What is she hearing that makes her smile and sway?
It begins to grow crowded. The slap of flipflops echoes off the hardwood floor.
There is a woman with a toucan-blue tuft of hair.
Her hair (almost! so close!) matches her turquoise pants.
One pictures her at the mirror this morning, carefully matching one to the other.
We're closed.
Much love to everyone who passed through this room today.
Goodnight.
Good night.

The end of a shift is different each time but it is always the writer who is the last to leave. A collecting institution is fundamentally challenged when open instructions constitute the artwork. Each time the work is acted out in the galleries, it produces and generates an endless stream of different effects, experiences, and time-based narratives. The presence of these narrative effects in the exhibition is fleeting: They do not stay in the eye of the beholder, materializing in the form of text only to be substituted by other sentences, paragraphs, and statements. Does this temporary production eventually—maybe secretly—become part of what can considered a collectible object? There is no clear institutional answer yet, but in keeping with its own good and ethical practice, the museum looks at the artist's practice to inform its curatorial or institutional decisions, ideally in a continuous dialogue. In García's own practice, these instant texts often leave the museum and circulate as books and documents. They become part of her Joycean project to collaborate with her audience to describe the world. She signed my copy of her book *All the Stories* (2011) and the inscription wishes me "a lot of pleasure and endless reading."

> What distinguishes this day, from yesterday. Hello, hello, & welcome.
>   This is your space. Isn't it?
> Horizontal stripes are definitely in. And high leather boots.
> Shoes that scoot along like whispers.
> Black head phones, a red sweat shirt & red shoe laces.
> Pink. Someone is crazy about pink.
> And a yellow scarf hanging down to the thighs, slightly tied.
> Black, black, black: low boots, socks, skirt, sweater. A gray scarf tucked
>   around the neck.
> Bright red scarf, dark red leather purse, gold metal, etc.
> How did we both decide to wear jeans to the museum.
> A low surf like rumble from outside the gallery. What made us come
>   to the museum today?
> Casual my name is casual. Left hand in my pocket, sweater over my
>   right shoulder.
> The grey fur around the top of the tall boots. Is it real or . . . ?

The reader's "endless reading" corresponds to the museum's endless learning about the conditions of exhibiting contemporary art. A seminal experience for every museum curator or educator is to spend time in the galleries, only in this case it involves one's own presence either as writer or reader, while both roles are consciously performed on stage both with and for the public. The work changes according to the writers' (and readers') knowledge, practice, mood, or attitude. There may be a stream of daily changing volunteers, producing a fairly broad range of texts, or there may be a small set of trained performers whose writing gets more and more crafted and consistent the longer the exhibition runs.[12] For a museum curator whose interest is to engage people in a dialogue with art, this learning curve is an eye-opener: The experience of art, as controlled by the museum as it might appear, is an open field of conscious dialogue, subconscious gestures, nonverbal patterns, missed encounters, and a highly charged atmosphere of individual tactics used to deal with one's state of being inside an art space.

> The space is so very lonely.
> All this wonderful art hangs, waiting to be looked at.
> Canvas, Oil on top of canvas, in a frame, on a wall.
> a group enters.
> mid conversation.
> They are a group of attractive young women.
> They talk and walk around, some of them have umbrellas.
> One sits on the comfy bench and plays with her hair.
> ANother girl joins her.
> They sit together and disrobe.

Unlike in a traditional performance with a beginning and an end, this situation—always already present for visitors—eventually constitutes an archived text that is linear and yet strangely circular and fragmented, each time reintroducing the notion of a beginning and an end in text form only: "What distinguishes this day, from yesterday[?]" is indeed a question that one would not typically pose to an artwork (as if it were capable of giving an "answer") but rather to the museum's Visitor Services. A considerably different text is produced when the gallery is empty than when it is crowded. In the latter case, the writer might try to respond to a rising wave of potential "characters" by producing a hectic flurry of shorter and shorter text fragments—precisely what happened in my own engagement with the work as a writer. And then there are cases of resistance—such as those described earlier by the art critic Tess Thackara—when a dialogue occurs between the different perspectives of the writer and the visitor. On this note, a stubborn and immutable presence of an anonymous figure in the same location over a long time forced me as the writer to acknowledge this reaction to the conditions of the performance. The minute my narrative responded to the presence of stillness, the "character" decided to leave, prompting me to reflect on the loss of presence and then to change the subject. For once, this character's search for his author had concluded successfully. He was publicly acknowledged, but then again, on whom did that brief moment on the screen actually register? It was in all likelihood a very intimate exchange in public that did not find a larger audience, and yet it stayed with me even more intensely than the memory of the hand-written notes and drawings that I received from other audience members and participants in response to my writing.[13]

"Look, that is funny, she is writing about a woman in a pink coat just like me," said a woman in a pink coat during the opening of *Instant Narrative* at SFMOMA, oblivious to the irony of her statement. After all, one simply doesn't expect to be an agent in a publicly exhibited text. Visitors will often use an artwork on display in the galleries as a personal mise-en-scène for "selfies" that they can then distribute via social networks, but they don't typically pay an entrance fee expecting to themselves show up on the wall—a place that they assume is exclusively reserved for serious art. In fact, what is on the wall is not them*selves,* but rather a fictional representation of them in narrative form. Yet the difference between the two is ever so minimal—almost imperceptible.

> you are doing a magnificent job says the mother of the narrator in
>   typical fashion.
> she is a source of great support
> it's been a while since i have been the subject of your writing she
>   says.
> what is this asks a young woman
> she wants to know if this is really art
> she has three stars behind her left ear
> you're typing all that she says.
> that's be cool to get paid to do that she says
> the payment is part of the reward
> but not a big part
> the narrator is talking to her mother.
> now it's time to focus on the room again.

The room is the stage but the distinction between stage and audience has been abolished. Not surprisingly, this blurry line between the real and the fictional at times prompts observation anxiety and a critical rejection of surveillance in public, but by the same token it also generates ecstatic intensity and a desire to be seen in public. A few distinct behaviors emerge almost instantly: Some characters never realize what is going on and are inattentive to their surroundings, while others do realize what is going on but leave the stage immediately. Then there are those who start acting and improvising for the writer, and others who are intrigued but prefer to lurk at the perimeter of the observable space. Just outside of the writer's vision, they occupy once again the traditional role of audience. Yet by being outside the frame and observing, they also become potential writers, comparing their own mental notes to what becomes manifest in the room and on the screen. At this point, the situation reaches a level of complex interaction that is hardly observable by a single writer; the individual occupying the role can merely probe into the abundance of information given at any moment.

The casual beauty of the way certain persons stand & witness.
Something in the way we move to bear (bare) witness.
Is this the first time you found yourself on a Gallery wall??

Beyond the body-centered performances of the 1970s and the conceptual gestures of relational aesthetics, García's works offer a structured activity, an open situation (as Tino Sehgal would define it),[14] and the participation of a whole community of writers, performers, and amateurs. It is fascinating to not only imagine but actually realize and document the fact that, while the museum stages and carefully frames the conditions for experiencing works of art, so much else is going on. The "narrative" of the gallery is constantly conditioned but also countered and undermined by the social dynamics of relationships between family members, lovers, strangers, guards, and so on. The art is filtered and mediated by these variable and often imperceptible conditions. It is not only the type of art that is present in the museum that has an effect on viewers, but also who is with them, next to them, or even directly addressing them. These participatory works are again "located" in the museum gallery, but now go beyond the Brechtian notion of breaking down the fourth wall in an attempt to stage its own method of production. Stage and auditorium are not only inseparable, they extend beyond the public realm of exhibition spaces and easily affect events in adjacent or even remote areas—perhaps including the museum's storage or publication activities—or may simply have ripple effects across social networking sites. While García's work is clearly indebted to historic predecessors like Brecht, her approach is a contemporary engagement with generative models of production. Ultimately, *Instant Narrative* constitutes the generation of a text as a contemporary investigation into modes of performing art with the audience. This opposes the traditions and localities of public scribes, public speakers, theatrical stage performers, and so on, but embraces art history and the museum into an expanded context. We experience an open text and stage of a shared authorship that includes the psychologies, politics, and social interactions of the work's public without staking a larger claim on a political theater or a social function.

Yet this work is far from apolitical; artists like García are staging and exploring the politics of spectatorship and public agency as an artistic act. In this field of continuous production and dialogue, the acts of engagement with the public are unscripted but nevertheless produce traces and effects of different voices and poetics. Something is not only voiced, said, spoken, or written; it is also heard, commented on, acted upon, rewritten, and distributed. While inserted in real time, García's dialogues involve first and foremost a writer and a reader, and the continuity and fluidity of this performative situation produces experiences of hybrid realities and agencies. In this dialogue there is no last word, and yet any published text—like this essay, for instance—by definition cannot avoid reaching a closure. Learning from participatory art, however, would mean keeping it as open as possible.

She implores him to read.
He is off to another gallery. Then she smiles, content with what she has read, and leaves. The writer here is content that she is content and what else counts?

Notes

The inserts that are interspersed throughout the essay are all taken from archived performances of Dora García's *Instant Narrative* (2006) at ICA London (2008), SFMOMA (2012), and the University of Michigan Museum of Art (UMMA), Ann Arbor (2014). No spellcheck or further editorial process has been performed on the inserts. All grammatical and orthographical errors have been preserved.

1. For a more general introduction to performing and collecting participatory art, see Rudolf Frieling, "The Museum as Producer: Processing Art and Performing a Collection," in *New Collecting: Exhibiting and Audiences after New Media*

Art, ed. Beryl Graham (Farnham: Ashgate, 2014), 135–58 (an earlier version of this text was also published as "The More Things Change: The Museum as Producer and Performer" in *The Challenge of the Object*, proceedings of the 33rd Congress of the International Committee of the History of Art CIHA 2012, published by Germanisches Nationalmuseum, Nuremberg, 2013); see also Rudolf Frieling, "Participatory Art: Histories and Experiences of Display," in *A Companion to Digital Art*, ed. Christiane Paul (Oxford: Wiley & Sons, 2016), 226–45.

2. See their dialogue in Allan Kaprow and Robert Smithson, "What Is a Museum?" in *Arts Yearbook*, 1967, reprinted in *The Collected Writings*, ed. Jack Flam (Berkeley: University of California Press, 1996), 43–51 (revised and expanded edition of the *Writings of Robert Smithson*, ed. Nancy Holt [New York: New York University Press, 1979]).

3. "Dialogical art" is Grant Kester's definition of participatory art and "conversational art" is Homi K. Bhabha's term, while others focus on community-based, collaborative, participatory, and democratic notions of art. All of these terms connote a multiplicity of voices without being defined by a clear social or political agenda.

4. Claire Bishop, *Artificial Hells* (London: Verso, 2012), 93.

5. See http://www.medienkunstnetz.de/works/hello-kaprow/.

6. Rudolf Frieling, "Towards Participation in Art," in *The Art of Participation: 1950 to Now*, ed. Rudolf Frieling (London: Thames and Hudson, 2008), 36. This essay and the exhibition catalogue include a broader view of the concepts and histories within contemporary art.

7. See http://www.thebeggarsopera.org/.

8. See the artist's statement in the chapter "The Kingdom Is Not the Museum," part of the online menu "Doors to the Kingdom" at http://www.macba.es/theking dom/index2.html/.

9. Ibid.

10. From the artist's manual for *Instant Narrative:* "The ideal performer is someone with a relatively high education (this is, not too young, well prepared for observing and for writing), able to type fast and without much faults. . . . The performer describes and writes down everything that is going on in the 'Instant Narrative' exhibition space, especially how the public looks and what the public does. The sentences must be short and to the point, the aim being that the audience recognizes itself, each person individually or as a group, in the texts that appears on the screen: they must instantly realize that the text is about them and that it is being written there and then, and that the person who writes it is looking at them at that very moment. The descriptions do not only have to be factual . . . but they can risk conclusions. . . . The only limits to deduction and fiction are: respect for the public (no offensive remarks) and that the description stays close enough to the real so that they can recognize themselves." Dora García,

unpublished artist's manual for *Instant Narrative*, n.d., archive of the San Francisco Museum of Modern Art.

11. See Tess Thackara's account of her experience of *Instant Narrative*, published on *Open Space*, SFMOMA's blog, http://blog.sfmoma.org/2012/02/descriptive-acts-part-one/.

12. *Instant Narrative* was performed at the CGAC Centro Galego de Arte Contemporanea (2009) over four months with twenty performers; at the GfZK Gesellschaft für Zeigenössische Kunst Leipzig (2007) the work was up for two months with four performers. At SFMOMA, approximately a hundred performers took on single shifts, occasionally repeating a shift, over the course of two months (see http://franwork.wordpress.com/2012/03/07/an-account-of-being-a-participant-in-dora-garcias-instant-narrative/ and Tess Thackara's response on the SFMOMA blog [see footnote 12]). At UMMA (2014) a group of ten students performed for three months. For a first-person account of this experience in Michigan, see http://www.michiganquarterlyreview.com/2014/04/dora-garcias-instant-narrative/ and http://bonnieabrown.wordpress.com/instant-narrative-performer/. The work premiered at Gallery Michel Rein's booth at the Artissima Fair in Turin (2006). Other venues include de Singel, 2007; ICA London as part of the group show "Double Agent," which toured to Mead Gallery, Warwick Arts Center, and the Baltic Centre for Contemporary Art, Gateshead, 2008 (see also http://worldflapjackday.blogspot.com/2008/03/instant-narrative-by-dora-garcia.html); the Spanish Pavilion at the Venice Biennial, 2011; and at the art space L'Aubette 1928, Strasbourg, 2011.

13. Again, this provokes the question: What should be done with a growing physical archive of generative and participatory works? Should one exhibit these effects in conjunction with the next performance?

14. Tino Sehgal's performative works, or "situations" in the artist's terminology, are based on two important historical references: Bruce Nauman's dictum that whatever the artist does in the studio must be art—including simple repetitive acts—and the expansion of 1960s dance into sculptural forms and simple movements, as in the works of Yvonne Rainer, Trisha Brown, or Simone Forti. Sehgal's situations, though, specifically engage the museum gallery. Equally important, his condition for selling a "situation" to a museum is void of any traceable written statement. The only proof that a work of art has been sold is a bank transfer and the oral accounts of the agents of this process.

CHAPTER 14

The Formation of New Reason

Seven Proposals for the Renewal of Education

Bernard Stiegler
Translated by Daniel Ross

PRIOR TO ANY QUESTION about the future of schooling, we must posit as a *principle* that only if we reflect on the university will it be possible to rethink primary and secondary education, and not the other way around; to do otherwise would be to say that we must conceive schooling on the basis of extracurricular "needs," whether these are economic, political, social, or even religious needs, and that universities should be required to train teachers according to these nonacademic imperatives.

Universities can train good teachers only by being oriented toward their primary purpose, which is research training, that is, *heuristic experience*. Train good teachers — that is, teachers who fear neither knowledge nor nonknowledge, knowledge of the latter being the most rigorous experience.

This applies equally to all the other tasks of professionalization that the university must undertake. When universities train professionals — whether in education or for any other profession — it is first and foremost to train and form rational minds by exposing them to heuristic experience, that is, critical experience.

As for vocational courses that do not require pursuit of the heuristic and critical aims that constitute the beginning and end of all academic life, they are catered for in so-called professional or vocational schools.

· · ·

Universities are intended to form and train students who have been initiated — even if only a little — into the heuristic mindset embodied in rational knowledge, and in order that some of them might themselves become researchers.

It is only insofar as this heuristic and thereby critical spirit is cultivated — a mindset that understands that disputation is the dynamic force of all knowledge—that the university can form good teachers, and good professionals in the many other sectors of economic and social life where intellectual plasticity (which is very different from "adaptability"[1]) has become a key factor through which new skills and competences can be continually acquired. It is for this reason that universities are today involved in professional training in many domains.

Furthermore, the role of the university as a central "hearth" through which the heuristic approach is fostered means that it should be the matrix within which scholarly programs in general are elaborated. It is on the basis of research by leading laboratories in both the human and social sciences—just as in the physical sciences, life sciences, earth sciences, engineering sciences, and so on—that programs and materials for teaching the disciplines at the secondary level of education should be conceived and designed.

The university is a more recent incarnation of what was originally called the academy in the sense instituted by Plato, and this feeds into the "academy" in the French administrative sense, where it refers to regions of national education.

The academy in the Platonic sense is a polemic hearth, founded on disputation that is initially essentially philosophical, then scientific, and through which the disciplines have come to be formed. These disciplines constitute circuits of transindividuation[2] that are not just long, but infinitely long,[3] that is, based on *anamnesic* experience, which for Socrates, in the *Meno*, was the matrix of all true (that is, rational) knowledge.[4]

. . .

What we learn from Socrates is that rational experience is dialogical. In this dialogism, two noetic souls who conduct a dialogue in good faith can create anamnesic experiences. This means that they can rediscover *by themselves*—in thinking for themselves, and therefore *within themselves,* without receiving their knowledge from the outside—that which *founds* this knowledge as, precisely, rational.

An objection might be that not *all* knowledge is fashioned in this way, that grammar, for example, is not apodictic (demonstrative), and neither is history, or geography, or still less literature, which is inherently fictional. And yet, if we can call these disciplines *academic,* it is essentially because they are composed of communities of peers capable of rational *argument.* These communities are capable, in other words, of turning their grammatical, historical, geographical, or literary experiences into *logical* experiences that are heuristic, critical, and rational in the sense that they are exposed to *public debate within the academic community.*

In the field of grammar, a debate about how best to describe the fact of great linguistic variability, through an economical and well-formed rule, is a rational debate. It is rational in the sense that it requires those who take part in this debate to possess the capacity for the kind of anamnesic experience that Socrates discusses with Meno. Hence, even if grammatical knowledge is not founded on anamneses of this kind, this does not prevent the ideality that grammar constitutes (what grammar tends to produce are ideal descriptions of the idioms that it formalizes) from taking geometrical idealization as canonical.[5]

Indeed, we posit in principle that this way of grammaticalizing *more geometrico* is not possible for "grammatical science," and on this point we fundamentally depart from the cognitivist theory of language inspired by the generative grammars of Noam Chomsky.[6] The fact remains, however, that grammar is an *academic* discipline to the strict extent that it does take geometrical rigor as its canon.

In the vocabulary of Gilbert Simondon, we can say that each of the protagonists of a dialogical scene is a psychic individual who, in dialogue with another psychic individual, coindividuates with them, that is, transforms with them—so long as the dialogue is fruitful, that is, provided that they each *learn* something from one another.

When this coindividuation leads to anamnesis, that is, to rational heuristic experience—which, precisely because it is rational, can be *shared* only by those who together create this experience (wherein a universal forms and imposes itself)—this coindividuation tends to produce the *transindividual,* that is, a meaning that can potentially be shared by everyone, beyond the limited dialogical field of two protagonists.

This dialogical becoming *beyond the scene of the dialogue* (as its becoming-public and its posterity) is a *process of transindividuation*—wherein circuits of transindividuation form, and where writing makes it possible to continue, exceed, and complete this dialogical scene. This theme was explored by Seneca in connection with epistolary exchange, and then by Foucault reading Seneca and writing about his reading, that is, in dialogue with Seneca, beyond that primary dialogical scene represented by Socrates.

That this possibility of exceeding the dialogical scene is its most intimate reality, concealed within it and as its very origin, and that this intimate and original reality of dialogue is revealed by the epistolary circuit organizing the time frame (the delay, the time of reflection, through which I become an other) within which traces circulate, and as such form circuits—all this can be found contained in a remark by Goethe:

> If we happen, under certain circumstances, to have written and sealed and dispatched a letter to a friend, which, however, does not find him, but is brought back to us, and we open it at the distance of some considerable time, a singular emotion is produced in us, on breaking up our own seal, and conversing with our altered self as with a third person.[7]

What we learn from Simondon is that there is no genuine psychic individuation that does not lead to a collective individuation. The horizon of this noetic collectivity consists in the possibility of inscribing, in one or another technical support, the traces of processes of coindividuation,[8] and· as such consists in the possibility of sharing, *beyond the dialogical scene* of this coindividuation, in a process of *trans*individuation that collective individuation constitutes.

Collective individuation is the metastabilization of transindividual fields—that is, fields of shared meanings. What is shared in this way need not always be rational. On the contrary, it is, perhaps, mostly irrational: Reason does not just arise spontaneously in the mind. If reason is initially animated as *desire* (as a reason to act for a potential to act), that is, as the *power to infinitize and idealize,* it nevertheless requires, as rational idealization, as the capacity to create apodictic experience from anamnesis (which is geometric anamnesis), a *formation or training* of this mind, of this spirit.

Such formation is the pathway and sometimes the pursuit (by researchers who make discoveries)[9] of a circuit of transindividuation within which an academic discipline is woven. This is why, if every mind or spirit is rational, the rationality that is effectively produced as a rational discipline requires the institution of a history of spirit, which needs to be *constantly reinstituted* by institutors, that is, by teachers.

That this historicity of mind and spirit could have led Hegel to the thought of absolute knowledge—which has itself been poorly understood and largely misinterpreted, perhaps and first by Hegel himself, for if with the title "absolute knowledge" Hegel described the conditions that have led to the proletarianization of all knowledge, that is, the *destruction* of all knowledge, this is certainly not how he himself understood the teleology of what he names Universal History[10]—should not mean that we become afraid to revisit the history of reason on the basis of a *pharmacological* history of processes of transindividuation. I will now explain this in more detail.

For, with respect to what is produced by academic institutions—from Plato's academy to the Académie de Versailles, and passing through the Académie française and academies of many other kinds, including those referred to by Kant as *Societäten der Wissenschaften,* scholarly or scientific societies, which emerge from the Republic of Letters and lie outside the university[11]—it is an *actualized* form of that reason contained as a potential in any psychic individual, which is established through a protean process of the transindividuation of rational minds, and as collective individuation.

These minds, which are literate, and which can therefore accede to the experience that Goethe described in his novel-of-formation *(Bildungsroman),* at various times have or create anamnesic experiences or critical experiences; that is, they take geometric anamnesis as their canon, of which these experiences are as such an *analogon.*

These rational minds—rational in the sense that they have *actualized à la lettre* their reason to act and their potential to act—also *transmit* their noetic experiences *à la lettre,* and thereby participate in the writing *à la lettre* of the collective individuation of their discipline. And they do so through their experience of (school)masters, who owe their title of "master" to the fact it has been conferred upon them by an academy through the process of rational writing or rational reading (the process of transmission) in which the transindividuation of their discipline essentially consists.

. . .

Socrates posed the question of anamnesis during a period when, according to both Socrates and his student Plato, anamnesic experience was being undermined by the practices of the Sophists. The latter, by taking advantage of the power to manipulate minds that the technique of writing confers on those who possess it, *give the feeling of knowing* to those they in fact *displace* from any experience of real knowledge—that is, anamnesic knowledge: If anamnesis is the experience of thinking for oneself, sophism is, on the contrary, that which *prevents* thinking for oneself, and therefore prevents thinking in general. It is a form of *stupidity,* and a stupidity that is all the more frightening in that it *believes it knows,* wholly unaware of its stupidity.

The outcome of this eristic scene, wherein philosophy constitutes itself by opposing sophism, is that we find in Plato an opposition between anamnesis and hypomnesis, where anamnesis refers to living and transformative reminiscence, and hypomnesis to memory as passive reproduction supported by a mnemotechnique—such as writing, which is, as Socrates explains, a *pharmakon.*

Writing is a *pharmakon* because—as the condition of the *polis,* both as sphere of public law (that is, law that is published and as such readable by all citizens *à la lettre*) and as sphere of rational debate (of *logos* as public logical exercise, in the *agora* as in the academy, a sphere itself *founded on the apodictic experience of geometry* that, as Husserl explained in the twentieth century, is inconceivable without "written . . . documentations" of the reasoning of geometers)[12]—it is a remedy. It is a remedy for the finitude of memory making it possible to infinitize, that is, to produce *infinitely long* circuits of transindividuation, through which geometers, for example, together share, through their geometric transindividuation, the experience of the *infinite* (always open) character of geometry itself.

But if it is a *pharmakon,* this is because it is not just curative but also poisonous—and it is for this reason that sophism can produce this logography that manipulates logical minds and thwarts, according to Socrates and his pupil, their rational dispositions, that is, their ability to think for themselves in the many fields of experience.

All this means that the condition of knowledge is pharmacological—as is the human condition in general, given that human life is the technical form of life, and given that all technics is pharmacological. Moreover, Jacques Derrida's reading of the *Phaedrus* in "Plato's Pharmacy" shows that anamnesic experience, that is, rational heuristic experience in general, *presupposes* hypomnesic experience—a theoretical and practical experience of exteriorized memory, that is, technicized memory.[13]

Contrary to what Plato would himself have us believe, the academy is above all a place where one practices this technical hypomnesis that is writing (and were it otherwise we would not even be aware of the *Dialogues*); it is a matter of acceding, *on the basis of hypomnesic practice,* to an *anamnesic experience*—even though Plato saw these two dimensions as forming an opposition.

Acceding to anamnesic experience on the basis of hypomnesic practice: What this means here is, first and foremost, acceding to anamnesis on the basis of a *capacity for reading and writing* without which, for example, it would not be possible to be a geometer. Nor, therefore, would it be possible to truly be a citizen: Citizens are those capable of acceding to the law *à la lettre,* and through that of participating in its writing, and therefore of individuating the law by individuating themselves. Citizens are, therefore, also those who have the capacity for an experience of political truth, that is, of justice, and of which geometrical truth would be the canonical model, that is, the ideal.

. . .

Anamnesis and hypomnesis, far from being opposed, are poles of a composition of mind (or spirit) with finitude, and all rational experience is subject to this pharmacological condition: Rational experience constitutes an attentional form of a very specific kind, in which the attention that I grant to an object leads to the universalization of the object.

Attention is a combination of retentions and protentions. Three categories of retention must be distinguished: *Primary* retention is what perception retains of the perceived in the *necessarily temporal* course of any experience whatsoever. *Secondary* retention is the transformation of a primary retentional process, that is, a perceptual process, into memory, that is, into the past, where what was initially perceived in the course of experience becomes past experience, and constitutes a competence for the one who retains this past. It is on the basis of their past (consisting of secondary retentions) that somebody who lives an experience retains (as primary retentions) what they are perceiving. Two individuals never perceive the same thing in the same object, and for this very reason (because their pasts are different)—except in the *scientific protocols of observation* that aim to *neutralize* everything that, in past experience, has not been *scientifically* transindividuated by the theory on the basis of which the observational protocols have been formulated, that is, to eliminate everything that is therefore not *shared* by the *peers* of that discipline (and it is called a *discipline* to the extent that this is true).

A primary retentional process always engenders a *protentional* process, that is, an *expectation* and anticipation in relation to the continuation of the course of what transpires, and which is retained primarily on the basis of the secondary retentions that underpin experience as its memory, that is, as the apprenticeship of experience, as lessons learned.

Tertiary retentions, however, also exist; these allow what occurs to be *retained outside oneself*, for example, as set down on an agenda, or in a diary or a letter to a friend. Tertiary retention refers to anything that, by preserving the trace of a lived experience, spatializes this lived past in a form that is not psychic, but technical. Hence retentions of this kind make it possible, for example, to overcome the "retentional finitude" of the geometer so that he can thus accede to geometric experience as such—that is, to anamnesis.[14]

. . .

It is *literal*, lettered tertiary retention that makes geometrical anamnesis possible. This is, at least, what Husserl argues in relation to the origin of geometry. And this literal tertiary retention was also the condition of the political community that the *polis* founded on the basis of the written publication of its law.[15]

This is so because tertiary retention conditions the possible arrangements of primary retentions and secondary retentions in a way that allows the projection of those protentions in which reasoning consists. And we call *rational* any form of attention to an object in which the protentions that it generates are rigorously compatible with the retentions from which they originate. This is what Aristotle described as the principle of noncontradiction.

An object of *logos* is an arrangement between primary retentions and secondary retentions that projects protentions by relying on the support of tertiary retentions that *determine* the *terms* of a process of transindividuation. The *psychic secondary retentions* of a geometer, for example, can become the psychic secondary retentions of *all* geometers; secondary retentions can become *collective*, and not just psychic, to the extent that they have been tertiarized, that is, *spatialized*.

It is the spatialization *à la lettre* of individual and collective secondary temporal retentions that constitutes *logos* as that specific attentional form that arose in ancient Greece.[16] *We ourselves*, however, are living through an epoch that is dominated by *new, industrial forms of tertiary retention*, new forms of the spatialization of the time of consciousness, or of the harnessing of the time of consciousness by material and therefore spatialized attention-capturing systems. As a result, those kinds of reading and writing that emerged from the cultures of the past seem to be in irresistible decline, *as do all rational forms of transindividuation.*

Having arisen in the nineteenth century, analogue retentions (which are *hypomnemata* as well as *pharmaka*) in the twentieth century became (in the form of cinema, radio, and television) the supports of an industry for *capturing attention* that led to a *deformation of attention.* Attention was *subjected* to industrial temporal objects whose purpose was to channel attention toward objects not of reason, nor even of desire, but *of consumption,* that is, *of the drives* (and of the drive *of destruction: consumere* means "to destroy").

Is the audiovisual and temporal form of analogue tertiary retention inevitably toxic, with no prospect of becoming a support for new forms of rational attention? Such a conclusion could be reached only by ignoring the role of cinema, not just as an art but as an instrument of ethnographic and anthropological investigation, but also of observation in countless fields of scientific imaging, or as an historical archive, that is, as a material of historical reason, and so on.[17]

Is it possible to imagine, then, a scholarly and academic practice of tertiary retention that would serve the formation and transmission of knowledge?

That the answer is clearly *yes* becomes obvious as the digitalization of the analogical brings such practices potentially within the reach of everyone. And the same is true for documentation technology, and for the capacity for the "bottom-up" production of metadata by indexation, that is, the possibility of creating new forms of metalanguage, and the possibility of forming those processes of transindividuation we refer to as *contributory,*[18] with "wikis" being a well-known example, thanks to Wikipedia.

Digital technologies (where the digital audiovisual is a new regime of what we still perceive analogically, just as when we read a digitalized text, such as a document that has been scanned, we still do so alphabetically) are contemporary forms of writing. They are forms of *electronic writing* circulating at the speed of light and utilized by young people on a massive scale, which makes them a key factor in an *essentially generational* socialization.

. . .

The *academic condition* (from kindergarten to the Collège de France) is *pharmacological,* and the role of academies in general is to bring about the *positivity* of the *pharmakon,* which is the only way of struggling against its toxicity.

This has always been the case, but with respect to the academic world today, something new must be built with the new kinds of *hypomnemata* and *pharmaka* that are the analogue and digital forms of writing.

Such an approach requires the constitution of a *pharmacology of spirit,* itself necessitating the constitution of a *general organology.* The goal of the latter is to approach psychosocial facts as processes of individuation that always involve organs of three kinds: the psychosomatic organs of the individual, the artificial organs that technical systems form, and the social organizations that are transindividuated through collective individuation.

In the contemporary situation, however, there is a specific problem that calls for new responses: Technological becoming has accelerated to the point that, today, *the university is structurally lagging behind technological development in general,* and the development of digital technologies in particular. The university *follows* technological developments, even though it supplies the scientific elements that make them possible. This situation is new, and it generates particularly virulent forms of toxicity.

In this context, what Joseph Schumpeter described as "creative destruction" is lived increasingly as sheer destruction, and leads increasingly often to the *destruction of both social and psychic structures.*[19] This acceleration is the result of the *economic war* being fought in an economic world that has itself been financialized, that is, made entirely subject to shareholders who subject the world to their requirements as mediated through marketing—which at the same time *short-circuits educational structures in general, from the parental sphere to school and university.*

This situation has become absolutely destructive, that is, irrational. Governments and public authorities must give time back to educators, and find new ways of regulating the media so as to reconstitute public affairs and create a new *res publica* (a new system of publication). They must create the conditions for new forms of research that in particular afford the younger generations—who are quicker to take hold of new forms of *pharmaka,* and the new processes of transindividuation emerging from them— the opportunity to participate, according to protocols yet to be developed, in contributory research activities capable of giving rise to theories and practices of *positive pharmacology.*[20]

Contributory research is an extension of Kurt Lewin's concept of action research. But it takes advantage of and builds upon this concept with the objective of reconstituting the relationship between the university, school, and what lies outside them, and of doing so via the digital *pharmakon*. It reopens the question of the relation between the university and the societies of amateurs that arose in the Republic of Letters, a question already posed by Immanuel Kant in *The Conflict of the Faculties,* but one that, in the epoch of digital networks and industrial technologies of contribution, resurfaces in a completely new way that the academic world can ignore only at the price of finding itself swept away forever.

. . .

Marketing truly has succeeded in *inverting intergenerational relations:* According to a 2011 French survey, 43 percent of purchasing decisions are prescribed by children, who condition their parents after having themselves been conditioned by marketing. The technologies of spirit that are tertiary retentions function here as *technologies of stupidity,* causing *generational short circuits* that destroy intergenerational relations, that is, education in all its forms, given that intergenerational relations are the foundation of every form of knowledge transmission.

This is certainly not to deny, however, that with *new knowledge in the hands of new generations* it will be possible to struggle against this situation. On the contrary, it is a question of ensuring that a *new intergenerational contract* is forged in the academic world, by implementing appropriate goals, means, and protocols for research.

The renewal of schooling, which presupposes the renewal of the university, must lead to an intergenerational contract in which the authority of knowledge is reconstituted as the heuristic experience of nonknowledge and the trust that society has in itself, and where all this is transformed by adopting new tertiary retentions in a reasoned way, rather than adapting to them according to the dictates of marketing.[21]

Such would be a true *society of knowledge,* of learning and scholarship, one not reducible to an "information society" or a "knowledge industry" (where the latter realizes itself essentially as the proletarianization of symbolic, psychic, intellectual, and spiritual activity). It would be a society that resolutely and deliberately struggles (through deliberative processes of all kinds, and well beyond the academic sphere) against the reign of stupidity.

More generally, it is here that the condition of escape from the current crisis will be found, a crisis that results from the reign of stupidity imposed by the proletarianization of mind and spirit.

This requires, however, the development of a new publishing and editing sector (across all media) with a new mission, given to it by public powers (national but hopefully also, one day, European authorities), a mission that takes as its priority serving a new society of learning, and that is not subservient to the hegemonic dictates of marketing. This is in the interests of industry itself: The great publishers Fernand Nathan, Hachette, Armand Colin, and many others were the fortunate beneficiaries of the educational policies that formed the bedrock of the Third French Republic. What we need now is academic research that leads to the construction of a new system of editing and publishing—but this will not be possible without strong public will and clear-sighted policies.

. . .

All this leads me to formulate seven proposals.

We must rethink schooling on the basis of a redefinition of the role of the university, as a crucible enabling a reinvention of academic teaching and the academic institution (heir to what was accomplished on the basis of Plato). This will, in the coming years:

1. put organological and pharmacological questions at the heart of its work, general organology constituting the paradigm of a *transdisciplinary heuristic;*

2. make tertiary retention not only an object of study, but an object of *practice;*

3. set up around these two objectives a *new integrated system of primary, secondary, and tertiary education;*

4. be tightly articulated with the new publication system generated by digitalization, transforming public space, public time, and the public thing, the *publishing and audiovisual industries having been accordingly reoriented by national and international public powers in the course of negotiations conducted at the instigation of academic authorities;*

5. take up the *question of cosmopolitanism* in this new context, which will also be that of a *post-consumerist globality,* organizing *within university networks* the *relation of the universal to diversality;*[22]

6. initiate a *new critique of knowledge, become technoscientific knowledge,* that is, a *critique* (in the Kantian sense) *of industrial power as such;* and

7. implement, in order to accomplish this, new protocols for contributory research, *tightly connecting new scholarly and scientific associations to the academic world as a whole.*

Notes

1. See Victor Petit, "Adaptation/Adoption," in *Vocabulaire d'Ars Industrialis,* in *Pharmacologie du Front national,* ed. Bernard Stiegler (Paris: Flammarion, 2013), 371–73.

2. [Stiegler's notion of individuation is derived from, and a modification of, Gilbert Simondon's, and the idea of transindividuation is based on Simondon's idea of the transindividual. Transindividuation means that when one individual individuates themselves (for example, by speaking, expressing themselves by exteriorizing themselves, thereby inventing and producing themselves), others are also brought to speak, in their own singular ways, and yet in so doing they share and take into account the language of the first speaker, including by opposing them, thereby bringing about their own transformation and individuation. Transindividuation is thus what forms a "we" as a collectivity of singular "I"s, but crucially it always does so through a milieu, composed of signs, symbols, words, gestures, and all kinds of signifying objects. This milieu is itself individuated, that is, transformed, through the process of individuals and collectives individuating themselves *in* this milieu, and yet at the same time this symbolic milieu, as shared by all and exterior to all, is what allows for meanings to form, which are not fixed but which may be *relatively* stable (but this also means, relatively *un*stable, which Stiegler calls "metastability"). All the signs and symbols of the symbolic milieu amount to various kinds of objects, techniques, and technics of memory, and through transindividuation these can lead to (relatively) stabilized, collectively shared meanings and understandings, or to destabilizations that can cause the destruction of meaning or the formation of new knowledge and interpretations. This stabilization occurs not only through commonality of usage but through authorities (for example, academic authorities, but also shamans, priests, and so on) who are authorized, in one way or another, to make selections in and among processes of transindividuation. —Trans.]

3. This point has been developed in Bernard Stiegler, *States of Shock: Stupidity and Knowledge in the Twenty-First Century,* trans. Daniel Ross (Cambridge: Polity Press, 2015).

4. I have analyzed this passage in three lectures given at the École de philosophie d'Épineuil, on October 9, 2010, and December 4, 2010, available at http://www.pharmakon.fr.

5. In ancient Greece, "kanon" referred to the reed or cane used by a mason or builder as a "ruler"; it then took on the figurative and general meaning of "rule," "principle," or "model."

6. On this point, see Bernard Stiegler, "Technics of Decision: An Interview," *Angelaki* 8, no. 2 (2003): 164–65.

7. Johann Wolfgang von Goethe, *Wilhelm Meister's Apprenticeship and Travels*, trans. Thomas Carlyle (Philadelphia: Lea and Blanchard, 1840), 88.

8. Simondon indicates this when he states that the technical milieu is the condition of the transmission of the transindividual. See Gilbert Simondon, *L'Individuation psychique et collective* (Paris: Aubier, 2007), 263–69.

9. If a scientific discovery is not an invention, this is because, while new, it nevertheless seems *always* to have been: It seems to exceed time and space or the event and scene of its discovery, instead constituting what Husserl called omnitemporality. And this is precisely why Socrates referred to anamnesis. But only if we understand in what rational transindividuation consists are we able to recognize such a viewpoint without having recourse to a transcendental metaphysics. On omnitemporality, see Edmund Husserl, *Cartesian Meditations,* trans. Dorion Cairns (Dordrecht: Kluwer, 1995), 127.

10. And Marxists, who inherit his dialectic by materializing it, will have no more understanding than Hegel of what Hegel calls universal history.

11. Immanuel Kant, *The Conflict of the Faculties,* trans. Mary J. Gregor (Lincoln: University of Nebraska Press, 1979), 25.

12. Edmund Husserl, "The Origin of Geometry," in *The Crisis of European Sciences and Transcendental Phenomenology,* trans. David Carr (Evanston, Ill.: Northwestern University Press, 1970), 357.

13. Jacques Derrida, "Plato's Pharmacy," in *Dissemination,* trans. Barbara Johnson (Chicago: University of Chicago Press, 1981).

14. That is, the fact that his memory "fails," that it constantly betrays him, that he cannot preserve his step-by-step reasoning from one evening to the following morning, finding himself forced to start again from the beginning. See Edmund Husserl, "The Origin of Geometry." The expression "retentional finitude" is from Derrida's commentary on Husserl, in Jacques Derrida, *Edmund Husserl's Origin of Geometry: An Introduction,* trans. John P. Leavey Jr. (Lincoln: University of Nebraska Press, 1978), 57.

15. Jean-Pierre Vernant, *Myth and Thought among the Greeks,* trans. Janet Lloyd and Jeff Fort (New York: Zone Books, 2006).

16. To pursue this point further, and everything that follows from it, see also Bernard Stiegler, *States of Shock.*

17. See Sylvie Lindeperg, *Clio de 5 à 7* (Paris: CNRS, 2000).

18. See Stiegler, *States of Shock.*

19. Joseph A. Schumpeter, *Capitalism, Socialism and Democracy* (London: Allen & Unwin, 1976).

20. Bernard Stiegler, *What Makes Life Worth Living: On Pharmacology,* trans. Daniel Ross (Cambridge: Polity Press, 2013), 4.

21. [Stiegler makes a fundamental contrast between adapting and adopting. Adaptation means to "make apt," hence to adjust to, to conform, and is commonly related to Darwinian ideas of adapting *to* the environment rather than, as human beings do, changing the environment itself. Adoption, on the other hand, contains an element of choice, of *opting*, and for Stiegler the crucial point is that for human beings all individuation is a question of adoption, whether of a way of life, technology, an idea, and so on. Adoption is thus associated with individuation, whereas adaptation is associated with disindividuation. To adapt to a norm, for example, presumes the norm exists independently of the individual, whereas to *adopt* a norm implies it exists *only in being adopted*. Adaptation presumes a relationship between preexisting terms, whereas the terms of an adoption do not preexist their relationship. Stiegler's concept of adoption is thus implicitly and explicitly a critique of the ideology of adaptation exemplified in the notion that "there is no alternative." —Trans.]

22. Regarding *post-consumerist globality*, see "*Ars Industrialis*: 2010 Manifesto," in Bernard Stiegler, *The Re-Enchantment of the World: The Value of Spirit against Populism*, trans. Trevor Arthur (London: Bloomsbury, 2014), 17–28; regarding the *relation of the universal to diversality*, see Patrick Chamoiseau, *Écrire en pays dominé* (Paris: Gallimard, 2002).

CHAPTER 15

Zoom Pavilion

Rafael Lozano-Hemmer and Krzysztof Wodiczko

ZOOM PAVILION IS A LIVE AUGMENTED-REALITY installation where people's presence in the public space is detected and projected onto the very ground where they are standing, using the world's brightest projectors. The piece is at once an experimental platform for self-representation and a giant urban microscope to connect the public to each other and to their city. The installation consists of two cranes with boom-lift projectors, cameras, and illuminators reaching a height of 18 meters. Under the crane-booms are four interactive projection areas on the ground, each measuring 22 by 16 meters, for a total coverage area of 1,428 square meters and 120,000 ANSI lumens of bright light.

As people walk into the illuminated interactive area, they are detected by custom-made computerized tracking systems which determine their position, velocity, and acceleration. Crucially, the system tracks not individuals but any group of people assembling in the space. Their image is immediately projected right beside them on the ground at a normal scale of 1:1. Four independent robotic cameras then zoom in to amplify the images with up to 35x magnification. The zooming sequences in and out of the public are disorienting as they change the entire image "landscape" from easily recognizable wide shots of the crowd to abstract close-ups of someone's hand, for example. The whole installation is in a constant fluid state of camera movement, highlighting different participants and creating a constantly changing animation based on optical amplification and tracking.

The piece is designed to be as inconspicuous as possible during the daytime: It has a minimal footprint so that normal emergency vehicle routes and bike and pedestrian traffic are completely unobstructed. At night the piece is very bright, but it still needs some of the public lighting

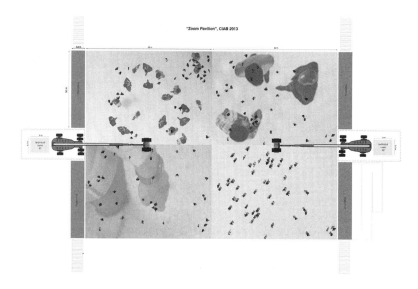

Figure 15.1. Rafael Lozano-Hemmer and Krzysztof Wodiczko, Zoom Pavilion, 2013 (4 views). Unrealized project conceived for the Fifth China International Architectural Biennial.

around the area to be turned off or masked. The installation is completely silent so it is not bound to disturb any neighboring activities.

Looking at the site for the Fifth China International Architectural Biennial, the artists decided that the most suitable approach would be to make interactive nighttime projections right onto the ground. Conceptually, transforming the ground is effective because it is the baseline for public space and architectural experience. Logistically, the Olympic Park has ample flat, evenly colored and textured ground surfaces, which make for great projection surfaces. The most important factor for the placement is to choose a location where there is already considerable foot traffic at night, so that people "encounter" the piece rather than having to walk far away specifically to see it.

Zoom Pavilion marks the first collaboration of artists Rafael Lozano-Hemmer and Krzysztof Wodiczko. Their practice often involves transformation of an existing built environment using projection technologies to "augment" the site with alternative histories, connections, or public relationships. The term "projection mapping" is now used often to describe techniques that Wodiczko was already deploying over thirty years ago. Meanwhile Lozano-Hemmer's contribution to the field over the past twenty years has been to develop intuitive ways to make mapped projections interactive with the general public.

Acknowledgments

This volume builds on activities and exchanges held during "The Participatory Condition," an international colloquium hosted by Media@ McGill at the Musée d'art contemporain de Montréal on November 15 and 16, 2013 (www.pcond.ca). We are grateful to everyone who presented at the conference: our keynotes, N. Katherine Hayles, Rafael Lozano-Hemmer, and Bernard Stiegler, as well as our sixteen panelists, Bart Cammaerts, Nico Carpentier, Julie E. Cohen, Mia Consalvo, Kate Crawford, Christina Dunbar-Hester, Rudolf Frieling, Birgitta Jónsdóttir, Jason Edward Lewis, Geert Lovink, Graham Pullin, Trebor Scholz, Skawennati, Christopher Soghoian, T. L. Taylor, and Jillian C. York. We warmly thank all colloquium participants for their presentations and for the discussions that ensued, which directly nourished the development of this book. Special thanks as well to François Letourneux, who ensured the smooth running of our collaboration with the Musée d'art contemporain de Montréal for this event.

Alongside the conference, Media@McGill offered a POOC (participatory open online course), facilitated by Alexandre Miltsov. Our colloquium research delegates, Alessandro Delfanti, Ian Kalman, Gretchen King, Adair Rounthwaite, and Erandy Vergara-Vargas, each assembled small teams of student and postdoctoral researchers. Together, they conducted background research on the topics covered in our panels, actively participated in the POOC, and posed the first formal questions to panelists during Q&A sessions with the public. We are grateful to all of them for their personal investment in this project and their contributions to its success. Some of the ideas first discussed in the POOC and with the delegates have made it into the volume. The complete contents of this online course are archived at www.pcond.ca/pooc.

This publication is made possible thanks to in-kind and financial support from Media@McGill, a hub of research, scholarship, and public outreach on issues and controversies in media, technology, and culture. Media@McGill's activities are primarily supported by a generous grant from the Beaverbrook Canadian Foundation. Our thanks go to Media@ McGill's administrative staff, Sophie Toupin and Mary Chin, for their assistance on this project, as well as to our copy editor, Matthew Goerzen, for his careful review of the original manuscript, and to our indexer, Jillian O'Connor. The editors would also like to thank the Social Science and Humanities Research Council of Canada and the Fonds de recherche du Québec—Société et Culture (FRQSC) whose funds directly and indirectly supported the colloquium and publication. In particular, we acknowledge the FRQSC's support for the Mediatopia research team undertaking the "Aesthetics, New Media, and the (Re)configuration of Public Space" research project.

We are very grateful to our anonymous reviewers and to the staff at the University of Minnesota Press—in particular, humanities editor Danielle Kasprzak and editorial assistant Anne Carter—for their careful guidance and feedback throughout the publication process, and to Michael Bohrer-Clancy, as well as Maura Neville at Westchester Publishing Services.

Finally, the editors would like to thank one another. Sometimes our great plans worked; sometimes we muddled through. At one time or another, each of us picked up the slack for someone else, and everyone brought unique perspectives and contributions to the table. It is rare to have so many great colleagues in a single department, and this project was a special opportunity to work together on ideas, rather than just the business of faculty governance. In the process, we also learned a great deal from one another.

Contributors

MARK ANDREJEVIC is a researcher at Monash University and a member of the Department of Media Studies at Pomona College. He is the author of *Reality TV: The Work of Being Watched*, *iSpy: Surveillance and Power in the Interactive Era*, and *Infoglut: How Too Much Information Is Changing the Way We Think and Know*.

DARIN BARNEY is Grierson Chair in Technology and Citizenship and associate professor in the Department of Art History and Communication Studies at McGill University. He is the author of *Communication Technology: The Canadian Democratic Audit*, *The Network Society*, and *Prometheus Wired: The Hope for Democracy in the Age of Network Technology*. He is coeditor with Andrew Feenberg of *Community in the Digital Age: Philosophy and Practice*.

BART CAMMAERTS is associate professor in the Department of Media and Communications at the London School of Economics and Political Science. His most recent books include *Mediation and Protest Movements* (coedited with Alice Matoni and Patrick McCurdy), *Media Agoras: Democracy, Diversity, and Communication* (coedited with Iñaki Garcia-Blanco and Sofie Van Bauwel), *Internet-Mediated Participation beyond the Nation State*, and *Understanding Alternative Media* (coauthored with Olga Bailey and Nico Carpentier).

NICO CARPENTIER is associate professor in the Communication Studies Department of the Vrije Universiteit Brussel (Free University of Brussels) and docent at Charles University in Prague. He is a research fellow at Loughborough University and the Cyprus University of Technology. He is an executive board member of the International Association for Media and Communication Research.

JULIE E. COHEN is the Mark Claster Mamolen Professor of Law and Technology at the Georgetown University Law Center. She is the author of *Configuring the Networked Self: Law, Code, and the Play of Everyday Practice* and *Copyright in a Global Information Economy*. She is a member of the Advisory Board of the Electronic Privacy Information Center.

GABRIELLA COLEMAN holds the Wolfe Chair in Scientific and Technological Literacy at McGill University. She is the author of *Coding Freedom: The Ethics and Aesthetics of Hacking* and *Hacker, Hoaxer, Whistleblower, Spy: The Many Faces of Anonymous*.

KATE CRAWFORD is visiting professor at MIT's Center for Civic Media, principal researcher at Microsoft Research, senior fellow at NYU's Information Law Institute, and codirector of the Council for Big Data, Ethics, and Society.

ALESSANDRO DELFANTI is assistant professor of culture and new media at the University of Toronto. He is the author of *Biohackers: The Politics of Open Science*.

CHRISTINA DUNBAR-HESTER is assistant professor in the Annenberg School for Communication and Journalism at the University of Southern California, where she teaches media, technology, and culture. Her research centers on the politics of technology in activism, and she is the author of *Low Power to the People: Pirates, Protest, and Politics in FM Radio Activism*.

RUDOLF FRIELING is curator of media arts at the San Francisco Museum of Modern Art. His curatorial projects include the online archive *Media Art Net* and major survey shows such as *In Collaboration: Early Works from the Media Arts Collection, The Art of Participation: 1950 to Now*, and *Stage Presence: Theatricality in Art and Media*. He is adjunct professor at the California College of Arts and the San Francisco Art Institute.

SALVATORE IACONESI is a robotics engineer, artist, and designer. He teaches digital design at La Sapienza University of Rome and at ISIA Design Florence. He is a TED Fellow, Eisenhower Fellow, and Yale World Fellow.

JASON EDWARD LEWIS is a Trudeau Fellow and the Concordia University Research Chair in Computational Media and the Indigenous Future Imaginary. He runs Obx Labs and codirects the Aboriginal Territories in Cyberspace research network as well as the Skins Workshops on Aboriginal Storytelling and Digital Media. In 2014, he was awarded the inaugural Best Electronic Literature prize from the Electronic Literature Organization.

RAFAEL LOZANO-HEMMER is an electronic artist who develops interactive installations at the intersection of architecture and performance art.

GRAHAM PULLIN is senior lecturer in interaction design and product design at DJCAD, University of Dundee. He wrote *Design Meets Disability*.

CHRISTINE ROSS is professor and James McGill Chair in Contemporary Art History in the Department of Art History and Communication Studies at McGill University, as well as director of Media@McGill, a hub of interdisciplinary research, scholarship, and public outreach on issues in media, technology, and culture. Her books include *The Past Is the Present; It's the Future Too: The Temporal Turn in Contemporary Art*, *The Aesthetics of Disengagement: Contemporary Art and Depression* (Minnesota, 2006), and *Images de surface: l'art video reconsidéré*.

TREBOR SCHOLZ is associate professor of culture and media, occasional artist, and chair of the Politics of Digital Culture conference series at The New School. He is the author of *Uberworked and Underpaid*, and, with Laura Y. Liu, *From Mobile Playgrounds to Sweatshop City*. He is the editor of many collections, including *The Internet as Playground and Factory*.

CAYLEY SOROCHAN is a doctoral candidate in communication studies at McGill University.

JONATHAN STERNE is professor and James McGill Chair in Culture and Technology in the Department of Art History and Communication Studies at McGill University. He is the author of *MP3: The Meaning of a Format* and *The Audible Past: Cultural Origins of Sound Reproduction*, and editor of *The Sound Studies Reader*. His website is at http://sterne works.org.

BERNARD STIEGLER is director of the Institute for Research and Innovation in Paris, a professorial fellow at Humboldt Universität in Berlin, and professor at the University of Technology of Compiègne. He is a cofounder of the Paris-based political group Ars industrialis. His books include three volumes of *La technique et le temps*, two volumes of *De la misère symbolique*, three volumes of *Mécréance et Discrédit*, and two volumes of *Constituer l'Europe*.

TAMAR TEMBECK is academic associate within Media@McGill, a hub of research on media, technology, and culture at McGill University. She is an art historian and curator, and the editor of *Auto/Pathographies* and coeditor of *Conflict[ed] Reporting*, a special issue of *Photography and Culture*.

KRZYSZTOF WODICZKO is professor of art, design, and the public domain at the Harvard Graduate School of Design. His major books include *Abolition of War, Krzysztof Wodiczko*, and *Critical Vehicles*. He won the Hiroshima Art Prize in 1998 for his contribution as an international artist to world peace, and he represented Canada and Poland in the Venice Biennale.

JILLIAN C. YORK is a writer and activist. She is the director for International Freedom of Expression at the Electronic Frontier Foundation, and her writing has been published by the *New York Times, Al Jazeera, The Atlantic, The Guardian, Al Akhbar English, Slate, Foreign Policy,* and *Die Zeit,* among other publications.

Index

AAC (Augmentative and alternative communication), 101–20

Aboriginal Territories in Cyberspace (AbTeC), xxix–xxx, 230, 232, 234, 240, 242–43

AbTeC. *See* Aboriginal Technologies in Cyberspace

Access, 7–8, 13, 23–24; to data, 125–27, 130, 136, 215; Internet, 44, 46, 82; to technology, 242. *See also* Open access

Aesthetic regime, xv–xvi, xxviii–xxix

Aesthetics, xxviii–xxix; participatory, xv, xvii; relational, xvi–xvii. *See also* Art, participatory

Agency: assemblage and, 196–99; digital culture and, 192–99, 215, 233; intention and, 198; political action and, 198–99; public, x, 266; tinkering and, 80, 85–86

Aisthesis, xxviii–xxix

Al Akhbar, 49

Al Arabiya, 45

Al Jazeera, 44, 45; whistle-blowing project, 48–49

Al Masry Al Youm, 49, 52

Althusser, Louis, ix–x

Amazon Mechanical Turk (AMT or MTurk), 61–62, 63, 69–70. *See also* Crowdsourcing; Digital labor; Turking

Anamenésis: experience, 270–71, 273; hypomnesis and, 274, 275; knowledge and, xxx

Antagonism, xvii, 10

Aouragh, Miriyam, 48

Applin, Sally, 177

Appropriate technology movement, 92–93

Arab Spring, 43, 48–51, 53; social media and, 50–51, 52–53

Arab world, 53; blogs, 45–46; censorship in, 43–44, 45–47, 53–55; independent media in, 50–53, 54; Internet and, 43, 44, 46; participatory media in, 43–55. *See also under names of specific countries*

Arendt, Hannah, xiii

Aristotle, 276

Arnstein, Sherry, 4, 5

Art, participatory, xv–xviii, xxix–xxx, 123, 251–66, 285; audience and, 264–65, 266; collectability of, 261–62; democratic, 253; interactive vs., xvii; museum and, 251–62, 265; passive, 258–59; political, xvii; post-Internet, xvi; purpose of, xxxivn25; theatre, 256–57. *See also* Aboriginal Territories in Cyberspace; *Instant Narrative*; La Cura; *Zoom Pavilion*

(*continued from page ii*)